REDUCTION AND PREDICTABILITY OF NATURAL DISASTERS

Proceedings of the Workshop "Reduction and Predictability of Natural Disasters" held January 5-9, 1994 in Santa Fe, New Mexico

Editors

John B. Rundle
Department of Geology Sciences &
Cooperative Institute for Research
in Environmental Sciences
University of Colorado

Donald L. Turcotte
Department of Geological Sciences
Cornell University

William Klein
Department of Physics and Polymer Center
Boston University

T0174025

Proceedings Volume XXV

Santa Fe Institute
Studies in the Sciences of Complexity

Routledge
Taylor & Francis Group
New York London

Director of Publications, Santa Fe Institute: *Ronda K. Butler-Villa*
Production, Santa Fe Institute: *Della L. Ulibarri*
Publications Secretary, Santa Fe Institute: *Marylee A. Thomson*

First published 1996 by Westview Press

Published 2018 by Routledge
711 Third Avenue, New York, NY 10017, USA
2 Park Square, Milton Park, Abingdon, Oxon OX14 4RN

Routledge is an imprint of the Taylor & Francis Group, an informa business

Copyright © 1996 Taylor & Francis

This volume was typeset using TEXtures on a Macintosh IIsi computer.

ISBN 13: 978-0-201-87049-7 (pbk)
ISBN 13: 978-0-201-87048-0 (hbk)

About the Santa Fe Institute

The *Santa Fe Institute* (SFI) is a multidisciplinary graduate research and teaching institution formed to nurture research on complex systems and their simpler elements. A private, independent institution, SFI was founded in 1984. Its primary concern is to focus the tools of traditional scientific disciplines and emerging new computer resources on the problems and opportunities that are involved in the multidisciplinary study of complex systems—those fundamental processes that shape almost every aspect of human life. Understanding complex systems is critical to realizing the full potential of science, and may be expected to yield enormous intellectual and practical benefits.

All titles from the *Santa Fe Institute Studies in the Sciences of Complexity* series will carry this imprint which is based on a Mimbres pottery design (circa A.D. 950–1150), drawn by Betsy Jones. The design was selected because the radiating feathers are evocative of the outreach of the Santa Fe Institute Program to many disciplines and institutions.

Santa Fe Institute
Studies in the Sciences of Complexity

Lectures Volumes

Vol.	Editor	Title
I	D. L. Stein	Lectures in the Sciences of Complexity, 1989
II	E. Jen	1989 Lectures in Complex Systems, 1990
III	L. Nadel & D. L. Stein	1990 Lectures in Complex Systems, 1991
IV	L. Nadel & D. L. Stein	1991 Lectures in Complex Systems, 1992
V	L. Nadel & D. L. Stein	1992 Lectures in Complex Systems, 1993
VI	L. Nadel & D. L. Stein	1993 Lectures in Complex Systems, 1995

Lecture Notes Volumes

Vol.	Author	Title
I	J. Hertz, A. Krogh, & R. Palmer	Introduction to the Theory of Neural Computation, 1990
II	G. Weisbuch	Complex Systems Dynamics, 1990
III	W. D. Stein & F. J. Varela	Thinking About Biology, 1993

Reference Volumes

Vol.	Author	Title
I	A. Wuensche & M. Lesser	The Global Dynamics of Cellular Automata: Attraction Fields of One-Dimensional Cellular Automata, 1992

Contributors to This Volume

Christopher Barton, U.S. Geological Survey, Center for Coastal and Marine Geology

Roger Bilham, University of Colorado

Jean Carlson, University of California, Santa Barbara

Stephen G. Eubank, Santa Fe Institute

C. Ferguson, Boston University

A. Gabrielov, Cornell University

Susanna Gross, University of Colorado

John P. Grotzinger, Massachusetts Institute of Technology

Vijay Gupta, University of Colorado

K. Haselton, Cornell University

Andreas Herz, University of Illinois

William Klein, Boston University

V. Kossobokov, Russian Academy of Sciences

William I. Newman, University of California, Los Angeles

Stuart Nishenko, U.S. Geological Survey, National Earthquake Information Center

S. L. Pepke, University of California, Santa Barbara

Daniel Rothman, Massachusetts Institute of Technology

John Rundle, University of Colorado

H. Saleur, University of Southern California

Charles Sammis, University of Southern California

Leonard Sander, University of Michigan

D. Sornette, CNRS, France

Brent Troutman, U.S. Geological Survey, Denver Federal Center

Donald Turcotte, Mathematical Institute, United Kingdom

Edward Waymire, Oregon State University

Contents

Preface

COMPLEX SYSTEMS IN THE EARTH SCIENCES:
AN INTRODUCTION TO PREDICTABILITY, MITIGATION, AND
REDUCTION OF NATURAL HAZARDS

Within the past five years, it has been recognized by the international community that it may be possible, through programs of systematic study, to devise
means to reduce and mitigate the occurrence of a variety of devastating natural
hazards. Among these disasters are earthquakes, volcanic eruptions, floods, and
landslides.[1,2,5,9] For example, it has recently been estimated that within 50 years,
more than a third of the worlds's population will live in seismically and volcani-
cally active zones.[4] The International Council of Scientific Unions, together with
UNESCO and the World Bank, have endorsed the 1990s as the International Decade
of Natural Disaster Reduction (IDNDR), and are planning a variety of programs
to address problems related to the predictability and mitigation of these disasters,
particularly in Third World countries. Parallel programs have begun in a number
of U.S. agencies, including the Department of Energy, the U.S. Geological Survey,
the National Science Foundation, and NOAA.

Clearly one of the most promising avenues of approach is to develop the capability to simulate these physical processes in the computer. Many of the recent

models are nonlinear in significant ways, including cellular automata-type or fractal growth models, and can thus be analyzed in a framework familiar to workers in complex system theory. Moreover, it is often the case that the occurrence frequency of disaster events generated by the models follow power laws, perhaps with cutoffs. Thus there is a spectrum of event sizes, from small to large, that are presumably related by the nonlinear dynamics of the process. The simulations can be used to study the fundamental physics of the process and, most importantly, to develop means to predict the patterns of occurrence of large events in the models using the statistics of small events. A variety of pattern recognition techniques may be useful here, including neural networks.[8,14] An open question is the extent to which the dynamics of the systems are determined by the physical structure, and vice versa. It may indeed be the case that physical structures such as faults, volcanoes, drainage networks and other landforms, and the dynamics by which they change, are linked by mutual feedback in a complex adaptive manner. Several areas of research that might benefit from the program outlined above include (1) earthquakes, (2) floods, (3) landslides, and (4) volcanic eruptions.

EARTHQUAKES

A large variety of nonlinear dynamical models for earthquakes have by now been studied.[6,7,10,12,15,20,21,24,25,26,27,30] At the present time, applications of statistical mechanical methods to the simulation and predictability of earthquakes seem to be enjoying great popularity. Most of these are based upon lattices of slider blocks, the nearest-neighbor version of which is the Burridge-Knopoff model. Two dimensional lattices of 10,000 to 20,000 blocks have been studied. Both deterministic and Monte Carlo models have been studied, and blocks with and without mass (inertia) have been used. Under a variety of conditions, one finds scaling in the frequency of events as a function of size. In some lattice models, the exponents evidently are those of a mean field spinodal, indicating the possibility of universality. Other models with different boundary conditions display period doubling, intermittancy, and chaotic bursts. The richness of model behavior, even from such a simple fault model, suggests that simulations can greatly aid our understanding of the dynamics of natural fault systems, which are in fact more generally characterized by fractal patterns of shear and tensional fractures. Prediction techniques using correlation dimension, phase portraits, and pattern recognition algorithms including neural networks should be tried with the goal of understanding and predicting the largest events on the basis of more frequent smaller events.

FLOODS

To a large extent, severe floods are often associated with the sudden development of large-scale drainage networks in response to unusually heavy precipitation.[13,16,19,22,23,28,29] Thus, the problem of the development of drainage networks bears importantly on the assessment of the hazard associated with a given topography and precipitation statistics. Recently, simulations of the development of drainage networks have been carried out using modified diffusion-limited aggregation models, and coupled nonlinear diffusion equations. Channel geometries are found that compare favorably to those observed in nature under a variety of conditions. Fractal exponents describing bifurcation ratios (~ 4) of channel generations and stream-length ratios (~ 2) are nearly identical to those of real stream channels. Assessment of the probability of flooding is associated with the peak discharge due to a given distribution of precipitation. Ideas can be tested and predictions developed using simulations like these carried out on a given topographic distribution. Cost effective strategies to mitigate severe floods can then be devised by altering the existing topography in specific ways (e.g., building dams and ditch networks).

LANDSLIDES

Landslides often occur in conjunction with severe precipitation on slopes that were previously unstable, or in association with ground shaking caused by earthquakes and volcanism. Details of the mechanisms are not well known; however, long runout landslides are evidently characterized by low internal friction and the tendency to flow out horizontal distances that are several times the original height of the sediment column. Submarine landslides can also occur, and may do considerable damage to undersea structures, including offshore oil platforms. Avalanches in mountainous regions may also cause considerable damage and loss of life. The recent interest in self-organized criticality and sandpile models is leading to a reexamination of these issues.[3,11,17,18] Study of particle avalanches on these lattice models are now well known to exhibit statistical behavior characteristic of both first- and second-order transitions. It seems likely that systematic analytical and numerical examination of these models may lead to substantial progress in understanding the fundamental physics. It may also be possible to predict the onset of large avalanches by the study of patterns associated with smaller events, using pattern recognition methods as mentioned above.

VOLCANIC ERUPTIONS

There are a number of significant hazards associated with large volcanic eruptions near populated and unpopulated regions. More than most hazards described above, effects can be regional and even global in extent, affecting the earth's climate for a period of years. This is the case with the 1991 Mt. Pinatubo eruption in which

4 km^3 of ash was injected into the atmosphere. Predicting the onset of the eruption and the extent of damage would allow timely evacuation of affected areas, warnings to aircraft that might fly through the ash plume, and may therefore substantially reduce risk of damage and loss of life. Many of the hazards associated with volcanic eruptions are the same as those associated with landslides,[3,11,17,18] as shown in the 1902 Mt Pelee eruption, 1980 Mt. St. Helens eruption, the 1985 Nevado del Ruiz eruption in which 23,000 people died, and the recent 1992 eruption of Mt. Spurr volcano in Alaska. In these events, the hazard was due either to a lahar (ash and mud) flow, a lateral blast and landslide, or a nuee ardante (gravity flow of extremely hot ash mixed with gas). In other words, the regional hazard can be viewed at least in part as a form of landslide, to which all of the comments above apply. Simulation techniques developed to understand landslides can thus be used to understand regional volcanic hazards. The global hazard relates to the effect of thermal and particulate perturbations on weather. This problem can be treated with GCM models, using the ash, gases, and thermal emissions as perturbations that may cause large effects due to nonlinear coupling terms in the hydrodynamic equations. Alternatively, it would be interesting to analyze phase portraits for global weather data both before and after large eruptions.

In order to address these issues, a program in reduction and mitigation of natural hazards has been established at the Santa Fe Institute. The first workshop in this series was held at SFI during January 1994, and attracted 25 scientists working in these areas. The second workshop will be held during June 1996 as part of the 18th Conference on Mathematical Geophysics. At the present time, plans call for publication of a number of volumes in the Santa Fe Institute series in the Sciences of Complexity that will serve to chronicle the scientific results of these meetings. This volume is the first in this series. We would especially like to thank the U.S. Department of Energy (DE-FG0394ER14425), Office of Basic Energy Sciences, Division of Geosciences; the National Science Foundation (EAR9318645); and the Geodynamics Program within the National Aeronautics and Space Administration (W91151) for generous support, without which this program and these volumes would not be possible.

REFERENCES

1. Advisory Committee on the International Decade for Natural Hazard Reduction. *Confronting Natural Disasters, an International Decade for Natural Hazard Reduction*. Washington, DC: National Academy Press, 1987.
2. Advisory Committee on the International Decade for Natural Hazard Reduction. *Reducing Disasters' Toll, The United States' Decade for Natural Disaster Reduction*. Washington, DC: National Academy Press, 1989.
3. Bak, P., C. Tang, and K. Wiesenfeld. *Phys. Rev. Lett.* **59** (1987): 381.

4. Bilham, R. *Nature* **336** (1988): 625.
5. Bryant, E. A. *Natural Hazards.* Cambridge: Cambridge University Press, 1991.
6. Burridge, R., and L. Knopoff. *Bull. Seism. Soc. Am.* **57** (1967): 341.
7. Carlson, J., and J. Langer. *Phys. Rev. Lett.* **62** (1989): 2632.
8. Casdagli, M., and S. Eubank. *Nonlinear Modeling and Forecasting.* Santa Fe Institute Studies in the Sciences of Complexity, Proceedings Volume XII. Reading, MA: Addison-Wesley, 1992.
9. Committee on Earth & Environmental Sciences. *Reducing the Impacts of Natural Hazards, a Strategy for the Nation.* U.S. Government Printing Office, 1992.
10. De Sousa Vieira, M., G. L. Vasconcelos, and S. R. Nagel. Preprint, 1993.
11. Feder, H. J., and J. Feder. *Phys. Rev. Lett.* **66** (1991): 2669.
12. Grassberger, P. Preprint, 1993.
13. Gupta, V. K., and E. Waymire. *J. Geophys. Res.* **95** (1990): 1999.
14. Hertz, J., A. Krogh, and R. G. Palmer. *Introduction to the Theory of Neural Computation.* Santa Fe Institute Studies in the Sciences of Complexity, Lecture Notes Volume I. Reading, MA: Addison-Wesley, 1991.
15. Huang, J., and D. L. Turcotte. *Geophys. Res. Lett.* **17** (1990): 223.
16. Hurst, H. E. *Am. Soc. Civil Eng. Trans.* **116** (1951): 770.
17. Hwa, T., and M. Karder. *Phys. Rev. Lett.* **62** (1989): 1813.
18. Kadanoff, L. P., S. R. Nagel, L. Wu, and S. M. Zhou. *Phys. Rev. A* **39** (1989): 6524.
19. Kramer, S., and M. Marder. *Phys. Rev. Lett.* **68** (1992): 205.
20. Lomnitz-Adler, J., L. Knopoff, and G. Martinez-Mekler. *Phys. Rev. A* **45** (1992): 2211–2221.
21. Narkounskaia, G., J. Huang, and D. L. Turcotte. *J. Stat. Phys.* **67** (1992): 1151.
22. Newman, W. I., and D. L. Turcotte. *Geophys. J. Intl.* **100** (1990): 433.
23. Rinaldo, A., I. Rodriguez-Iturbe, R. Rigon, E. Ijjasz-Vasquez, and R. L. Bras. *Phys. Rev. Lett.* **70** (1993): 822.
24. Rundle, J. B., and S. R. Brown. *J. Stat. Phys.* **65** (1991): 403.
25. Rundle, J., and D. Jackson. *Bull. Seism. Soc. Am.* **67** (1977): 1363.
26. Rundle, J. B., and W. Klein. *J. Stat. Phys.* (1993): in press.
27. Sornette, R., P. Davy, and A. Sornette. *J. Geophys. Res.* **95** (1990): 17353.
28. Turcotte, D. L. *Fractals and Chaos in Geology and Geophysics.* Cambridge: Cambridge University Press, 1992.
29. Turcotte, D. L., and L. Greene. *Stochastic Hydrol. Hydraul.* **7** (1993): 33.
30. Vallette, D. P., and J. Gollub. *Phys. Rev. A* (1993): in press.

Societal Effects of Natural Disasters

Stuart P. Nishenko† and Christopher C. Barton‡
†U.S. Geological Survey, National Earthquake Information Center, Denver, CO 80225
‡U. S. Geological Survey, Center for Coastal and Marine Geology, St. Petersburg, FL 33705

Scaling Laws for Natural Disaster Fatalities

Global comparisons of earthquake fatalities during the nineteenth and twentieth centuries and comparisons of fatalities from different types of disasters occurring in the United States during the twentieth century demonstrate that earthquakes and other natural disasters can be described with fractal or power-law size-frequency distributions. The introduction of a scaling exponent, D, provides an index to describe and compare losses associated with earthquakes and other natural disasters in space and time. The self-similar nature of these distributions permits the probability of infrequent, catastrophic events to be directly estimated from the rate of occurrence of smaller, more frequent disasters. Probabilistic estimates for the occurrence of catastrophic events provides a quantitative basis for prioritizing global disaster relief and mitigation programs and developing multidisaster mitigation programs at the national level.

Reduction & Predictability of Natural Disasters, Eds. Rundle, Turcotte, & Klein,
SFI Studies in the Sciences of Complexity, Vol. XXV, Addison-Wesley, 1996

INTRODUCTION

Earthquakes, hurricanes, tornadoes, and floods are complex natural phenomena that can be characterized by power-law size-frequency distributions.[2,7,12,21] Natural disasters, as defined by the loss of life and property, also exhibit power-law or fractal size-frequency distributions in which small disasters occur more frequently than large disasters. An understanding of how a particular disaster scales to other disasters caused by the same phenomenon and to disasters caused by other phenomena is fundamental to the development and evaluation of natural disaster mitigation and hazard reduction programs.

A scaling law is termed self-similar where the frequency-size distribution has no characteristic size or length scale. The power-law distribution is such a scale-invariant distribution. In addition, any power-law frequency-size distribution with a scaling exponent that is noninteger is considered to be fractal. A power, Pareto, or fractal distribution of random variables is self-similar and scaling in that the same distribution is obtained under truncation.[12] Earthquakes exhibit self-similar scaling behavior in space, time, and magnitude.[10,14,17] The Gutenberg-Richter (G-R) frequency-magnitude or b-value relationship,

$$\log N_c = a - bM \qquad (1)$$

where N_c is the cumulative number of events in a specified area and time interval greater than or equal to a particular magnitude, M [where M is proportional to $\log_{10}$ (seismic wave amplitude)], is an example of power-law scaling in seismology.[20] The constants a and b in the G-R relationship are determined from the rates of occurrence of smaller magnitude earthquakes and are used to extrapolate the rates of occurrence of infrequent larger magnitude events.

In this study, we investigate the utility of using the distribution of losses from past disasters to estimate future risk. Catalogs or inventories of natural disaster losses contain the integrated effects of fluctuations in the frequency of occurrence, slow vs. rapid onset time, small vs. large damage areas, and variations in damage intensity. This study presents three perspectives on losses due to natural disasters. The first is a global comparison of earthquake fatalities that have occurred during the twentieth century. The second is a comparison of earthquake fatalities in the same country during the nineteenth and twentieth centuries. The third comparison is of losses from different types of disasters that occurred in the United States during the twentieth century. We find that losses associated with earthquakes, hurricanes, floods, and tornadoes all exhibit power-law frequency-size distributions. This commonality provides a framework for the comparison of losses resulting from different types of disasters and allows use of the rate of occurrence of small disasters to infer the frequency of larger disasters.

GLOBAL EARTHQUAKE FATALITIES

Records of earthquake fatalities in China, Japan, Italy, Iran, Peru, Turkey, Chile, and India were examined to establish the equations that describe the frequency-size distribution or scaling behavior of earthquake disasters. These countries experienced the majority of earthquake fatalities during the twentieth century,[4] and provide a rich sample of geographic and tectonic regimes, types and sizes of earthquakes, construction techniques, and population densities. Estimates of both direct and indirect fatalities from individual earthquakes have been combined in our analysis.[5] Direct fatalities are those caused by the collapse of buildings and other structures during the earthquake itself. Indirect fatalities are those caused by secondary effects (e.g., landslides and tsunamis triggered by the earthquake) as well as fires, exposure to cold, epidemics, etc. In many cases it is impossible to distinguish the number of direct and indirect fatalities from available reports. For those events with more than one fatality estimate, we have bracketed the loss with both maximum and minimum estimates.

In the same way that earthquake magnitudes are proportional to $\log_{10}$(seismic wave amplitude)[9,16] or that the magnitude of a war or deadly quarrel can be defined as $\log_{10}$(number of fatalities),[15] we define the magnitude of the loss associated with a particular earthquake (or other natural disaster) as

$$F = \log_{10}(\text{number of fatalities}) \qquad (2a)$$

where F is the fatality magnitude. Variations in overall population size and rates of growth from country to country can be accounted for by dividing the number of fatalities for a given event by the population of the country at the time of the event. The normalized fatality magnitude, F', is then

$$F' = \log_{10}(\text{number of fatalities/population}) . \qquad (2b)$$

Earthquake fatality-frequency distributions for China, Japan, Italy, Iran, Peru, Turkey, Chile, and India are presented in Figure 1. These data have been normalized by the population of country at the time of the earthquake,[13] and represent a macrocomparison of earthquake fatalities. Once differences in population size are accounted for, 6 out of 8 countries [China, Japan, Italy, Chile, Peru, and India] exhibit linear fatality-frequency behavior in log-log space that extends over 3 to 5 orders of magnitude in loss. Least-squares fits to the central linear segments of the cumulative fatality-frequency distributions in Figure 1 are used to determine the power-law equation,

$$\log N_C = a - DF' \qquad (3)$$

where N_C is the cumulative number of events with normalized fatalities greater than or equal to F', the normalized fatality magnitude; D is the slope of the distribution

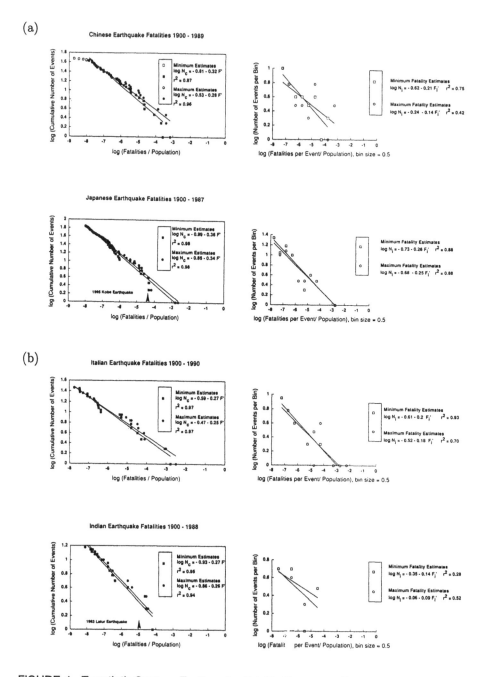

FIGURE 1 Twentieth Century Earthquake Fatality-Frequency Distributions. Cumulative (left) and interval (right) fatality-frequency distributions for China, Japan, Italy, (continued)

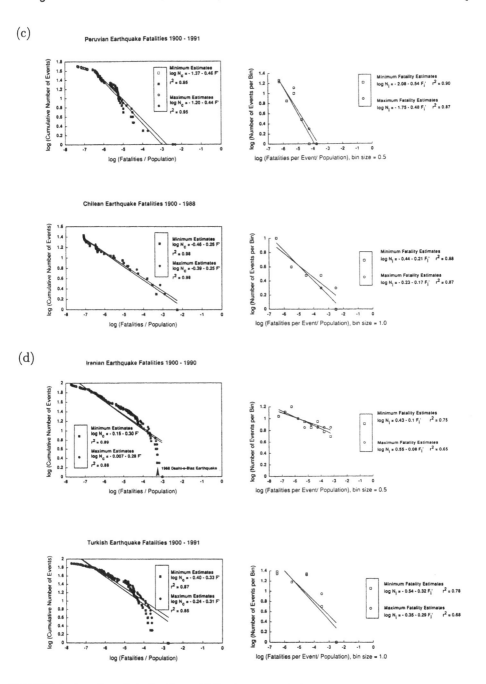

FIGURE 1   (continued) India, Peru, Chile, Iran, and Turkey illustrate a number of similarities after differences in population size are accounted for. Cumulative plots in (continued)

FIGURE 1   (continued) log-log space show the number of events of size F' or greater, where F' is the normalized fatality magnitude. Least-squares fits to the linear portion of the cumulative data (solid symbols) are shown for both minimum (squares) and maximum (circles) fatality estimates. Events with fatalities less than the threshold of completeness are shown as open symbols. Least-squares fits to the interval data are shown in the right hand column. The linear behavior of these distributions in log-log space over 3 to 5 orders of magnitude in loss illustrates the underlying power-law scaling behavior of earthquake fatalities in these diverse regions. The pronounced curvature or truncation of the fatality-frequency distribution for Iran and Turkey indicates the frequent occurrence of maximum fatality disasters in these two countries. Arrows show normalized fatality levels for the 1968 Dasht-e-Biaz, Iran; ; and 1995 Kobe, Japan earthquakes.

TABLE 1   Cumulative least-square fits to the twentieth century earthquake fatality-frequency data in Figure 1. N_c is the cumulative number of events with normalized fatalities greater than or equal to F'. F' is defined as log(fatalities/population).

Country	Minimum Fatality	Maximum Fatality
China	$\log N_c = -0.81 - 0.32F'$	$\log N_c = -0.53 - 0.28F'$
Japan	$\log N_c = -0.99 - 0.36F'$	$\log N_c = -0.85 - 0.34F'$
Italy	$\log N_c = -0.59 - 0.27F'$	$\log N_c = -0.47 - 0.25F'$
India	$\log N_c = -0.93 - 0.27F'$	$\log N_c = -0.86 - 0.26F'$
Peru	$\log N_c = -1.37 - 0.46F'$	$\log N_c = -1.20 - 0.44F'$
Chile	$\log N_c = -0.46 - 0.26F'$	$\log N_c = -0.39 - 0.25F'$
Iran	$\log N_c = -0.15 - 0.30F'$	$\log N_c = -0.007 - 0.28F'$
Turkey	$\log N_c = -0.40 - 0.33F'$	$\log N_c = -0.24 - 0.31F'$

or scaling exponent; and a is a constant. Both the cumulative and interval forms of the fatality-frequency distributions are plotted in Figure 1. While the cumulative form of Eq. (3) is commonly used in the literature, the interval form allows testing of the data for completeness in various size ranges or intervals. Various fits to these data including cumulative least-squares and interval least-squares were calculated to help constrain the value of D, the scaling exponent. The cumulative least-squares fits to both maximum and minimum sets of normalized earthquake fatality data are summarized in Table 1.

While the overall fatality-frequency behavior of the majority of countries is similar, there are some important differences. The departure from linearity (break in slope or roll-over) at the low fatality (left) end of the cumulative fatality-frequency plot for Chinese and Peruvian earthquakes in Figure 1 is interpreted to represent a deficit in the reporting of small fatality events. For earthquake magnitude-frequency distributions, this break in slope is related to the detection threshold of seismograph networks. Similarly the deficit in Figure 1 is related to the selection and reporting criteria (ten or more fatalities) of the earthquake fatality catalogs used on our analysis.[5] In contrast to China and Peru, fatality data from Japan, Italy, Chile, and India show little or no rollover at the low-fatality end during this same time period. Both the cumulative and interval forms of the frequency-size distribution are useful in determining where the threshold of complete reporting occurs. Knowledge of where this threshold occurs is critical when comparing data from different time periods or different regions. All of data sets in Figure 1 appear to be complete above an absolute threshold value of 10 fatalities per event.

In contrast to the majority of countries in Figure 1, both Iran and Turkey exhibit nonlinear cumulative fatality-frequency distributions in log-log space. Fatalities from both countries appear to terminate abruptly at normalized fatalities of $10^{-4} - 10^{-3}$ [between 10,000 and 100,000 fatalities per event ($F = 4$ to 5)]. We suggest that this abrupt or nonlinear decay of the cumulative number of events is related to the size of the maximum disaster in Turkey or Iran. It has been long recognized that earthquakes have an inherent finite size or maximum magnitude.[11] The Gutenberg-Richter recurrence relationship does not extend to infinitely large magnitudes but has a limiting size, which is dictated by the local geology and tectonics. Similarly, it can be argued that there are a maximum number of fatalities that can be caused by a single natural disaster. At the extreme, this would be the entire population of the region at risk. The 1968 M 7.1 Dasht-e-Biaz, Iran earthquake, for example, totally destroyed all of the buildings in the village of Dasht-e-Biaz, killing 74% of the local inhabitants[1] (or 0.06% of the total Iranian population at the time of the earthquake, see arrow in Figure 1). Both Iran and Turkey have long histories of catastrophic earthquakes involving large numbers of fatalities. Typical construction materials used in these countries [mud wall/adobe brick, rubble masonry] are very vulnerable to earthquake damage and are highly lethal.[1] The fatality distributions in Figure 1 indicate that poor construction has pushed societal fragility to the maximum limit in Iran and Turkey during this century.

TEMPORAL VARIATIONS OF EARTHQUAKE FATALITIES

Historic (pre-twentieth century) fatality data provide a long-term perspective on the impact of natural disasters on society. This earlier time period predates the development of many hazard mitigation programs, and provides a baseline for comparison with twentieth-century disasters.

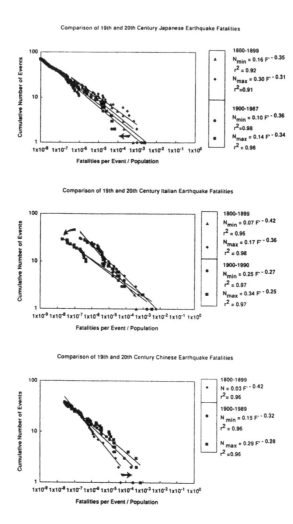

FIGURE 2 Comparison of nineteenth- and twentieth-century earthquake fatalities. Earthquake fatality-frequency distributions for nineteenth- and twentieth-century Japan, Italy, and China demonstrate how losses have changed as a function of time. All data have been corrected for differences in population at the time of the individual events, and have been truncated at the threshold of completeness. Nineteenth-century maximum and minimum fatality estimates are shown as diamonds and triangles, twentieth-century maximum and minimum fatality estimates are shown as squares and circles, respectively. While the slope of the Japanese fatality-frequency curve has remained constant with time, the number of fatalities associated with the largest disasters has decreased during the twentieth century. In Italy, there have been fewer small fatality events during the twentieth century, hence a flatter slope and smaller D value. The number of high-fatality Italian disasters ($> 5 \times 10^{-6}$), however, (continued)

FIGURE 2 (continued) has remained approximately constant during these two periods. For China, nineteenth-century minimum and maximum fatality estimates are essentially identical, and only one set is shown for comparison (solid diamonds). Overall, there were more large fatality events during the twentieth century in China, and the slope of the fatality-frequency curve decreased from 0.42 to 0.28 and 0.32.

Exactly how the scaling of fatalities changes as a function of time varies by region. Earthquake disaster histories for China, Japan, and Italy illustrate three types of behavior (Figure 2). In Japan the scaling exponent, D, appears to have remained constant from the nineteenth to the twentieth century (0.33 vs. 0.35, respectively). There is a suggestion, however, that the number of fatalities associated with the largest disasters decreased during the twentieth century. For China and Italy, decreases in the value of the scaling exponent, D, from the nineteenth to the twentieth century are significant at the $\alpha = 0.05$ level (Figure 2). For Italy, the overall decrease in slope $(0.42 - 0.36$ to $0.27 - 0.25)$ reflects a decrease in the number of small fatality events during the twentieth century. For more catastrophic earthquakes with normalized fatalities greater than 10^{-5}, both the nineteenth- and twentieth-century distributions of Italian earthquake fatalities are essentially identical. For China, nineteenth-century minimum and maximum fatality estimates are essentially identical, and only one set is shown for comparison in Figure 2. In contrast to Italy, however, the overall decrease in D for China from the nineteenth to the twentieth century (0.42 to 0.28) is caused by an increase in the number of high-fatality earthquakes during the twentieth century.

FUTURE EARTHQUAKE FATALITIES

Fatalities associated with the 1993 Latur, India (9743 dead) and 1995 Kobe, Japan (5281 dead) earthquakes are a grim reminder of our continued vulnerability to natural disasters. The arrows in Figure 1 compare the normalized fatality levels for these earthquakes with past disasters and suggest that these recent disasters were extreme, infrequent events. Projections of the rates of urbanization near earthquake belts indicate that 290 million people worldwide will live in regions of seismic risk by the end of this century. Twice that many people are estimated to be at risk by 2035.[3] The geographic and temporal stability of the scaling relationships for earthquake fatalities in Figure 1 and 2 suggests the use of these distributions to estimate the size and frequency of future earthquake disasters. For countries with little or no disaster mitigation, these estimates provide projections of future losses for developing disaster relief and mitigation programs. For countries with existing natural disaster mitigation programs, these projections can provide a baseline for the evaluation of the effectiveness of current programs at a later date.

TABLE 2 Poisson probability estimates for the occurrence of an earthquake with greater than 10,000 fatalities (i.e., $F > 4$, where $F = \log$ [fatalities per event] during 1, 10, and 20 year exposure times. Return period estimates listed in the last column are based the normalized fatality-frequency distributions of earthquakes in each country during the twentieth century and are indexed to 1990 census values.[12]

| | Exposure Time | | | |
Country	1 year	10 years	20 years	Return Period, years
China	0.07	0.54	0.79	13
Japan	0.03	0.29	0.50	29
Italy	0.03	0.26	0.45	33
India	0.02	0.22	0.39	40
Peru	0.03	0.26	0.45	33
Chile	0.02	0.22	0.39	40
Iran	0.09	0.60	0.84	11
Turkey	0.08	0.57	0.81	12

Estimates of return periods for catastrophic $F > 4$ (10,000 fatality) earthquakes in each country are based on the cumulative fatality-frequency distributions in Figure 1, and are adjusted to 1990 census estimates.[22] These return period estimates are similar to those for 100-year floods (i.e., the largest flood that occurs once a century) and do not imply any periodicity in the time intervals between events. Table 2 lists return periods and the Poisson probabilities for the occurrence of an $F \geq 4$ disaster during exposure windows of 1, 10, and 20 years duration in each of the eight countries studied. Not surprisingly, China, Iran, and Turkey have the highest probabilities [and the shortest return times (11–13 years)] for catastrophic $F \geq 4$ events these during these three exposure periods. Chile and India have the longest return times (40 years) and the lowest probabilities, followed by Peru, Italy, and Japan (33 to 29 years).

NATURAL DISASTER FATALITIES IN THE UNITED STATES

Fatalities associated with other types of natural disasters can also be characterized by power-law size-frequency distributions. Figure 3 compares the cumulative distribution of earthquake, flood, hurricane, and tornado fatalities in the United States during the twentieth century. These fatalities have not been normalized by population size. Reports for specific types of disasters span different time periods, and these data have been normalized by the length of the reporting period to reflect the annual rate of occurrence.[6,8,18,19,23] As in the case of the earthquake fatalities, the roll-off at the low fatality (left) end of the cumulative flood data clearly illustrates the effects of incomplete or partial reporting. Annual reports of U.S. floods since 1986 itemize only those fatalities associated with major events, whereas fatalities associated with smaller events are reported only as part of the annual total.[23] In contrast, the sharp termination of the tornado frequency-fatality distribution at 10 fatalities per event reflects the minimum number of fatalities per event in our data base.[6]

Comparison of 20th Century United States Natural Disaster Fatalities

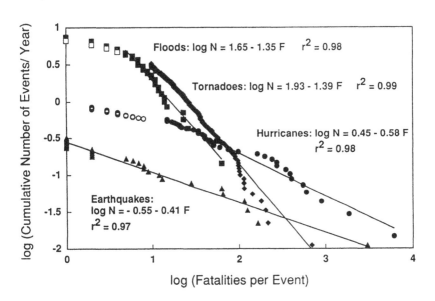

FIGURE 3   Comparison of Natural Disaster Fatalities in the United States, 1900–1990. Cumulative frequency-size distributions for annualized earthquake (triangles), flood (squares), hurricane (circles), and tornado (diamonds) fatalities. Solid symbols are those data above the threshold of complete reporting, and are used for the cumulative least-squares fits shown. Open symbols are those data below the threshold of complete reporting. These data group into two families that demonstrate the linear behavior over 2 to 3 orders of magnitude in loss. Earthquakes and hurricanes are low frequency, high maximum fatality disasters with shallow (D = 0.4 − 0.6) slopes; floods and tornadoes are high frequency, relatively low maximum fatality events and exhibit steeper (D = 1.4) slopes. The crossover point where fatalities from individual hurricanes dominate fatalities from individual floods and tornadoes occurs at approximately 100 fatalities per event.

The cumulative fatality-frequency curves in Figure 3 are linear over 2 to 4 orders of magnitude in loss and define two groups or families of fatal disasters. The first group, which includes both earthquakes and hurricanes, is characterized by a relatively flat slope ($D = 0.4 - 0.6$) and a number of large (> 100) fatality events during the twentieth century. At this scale, the primary difference between earthquake and hurricane fatalities during this time is the annual rate of activity. Fatal hurricanes occur more frequently than fatal earthquakes, however; the proportion of large to small events is similar in both cases. Flood and tornado disasters comprise the second type of fatality-frequency distribution, with steep slopes ($D = 1.3 - 1.4$), relatively few large fatality events, and higher annual rates of activity.

The integrated effects of slow- vs. rapid-onset times, small- vs. large-damage areas, and variations in damage density contained in the size-frequency distributions of this suite of disasters indicates that not all disasters have the same impact. Differences in the slope or scaling exponent of these distributions translate into significant differences in their impact at different fatality levels. On an annual basis, small fatality ($F = 1$) floods and tornadoes are 3 to 5 times more frequent than $F = 1$ hurricanes, and 18 to 30 times more frequent than $F = 1$ earthquakes. Large fatality events (i.e., $F = 2$ and larger) on the other hand, are less frequent and were predominantly related to hurricanes during the twentieth century. The crossover point where fatalities from individual hurricanes exceed those caused by floods and tornadoes occurs at approximately $F = 2$ (i.e., 100 fatalities per event). Based on the data in Figure 3, we have computed probabilities for the occurrence of natural disasters in the United States with $F = 1$ (10 fatalities per event) and $F = 3$ (1000 fatalities per event) during exposure windows of 1-, 10-, and 20-year duration. These estimates are presented in Table 3.

SUMMARY

Natural disasters are the product of complex interactions between nature and society. We have provided three different perspectives of life loss due to natural disasters to demonstrate how individual disasters scale with respect to other disasters caused by the same phenomenon and to disasters caused by other phenomena.

The first, a global comparison of earthquake fatalities, illustrates the basic similarity of fatality-frequency distributions among a number of geologically and societally diverse regions. Fatalities associated with earthquakes are described well by power-law size-frequency distributions. The introduction of a scaling exponent, D, provides an index to describe and compare losses associated with either earthquakes or other types of natural disasters in space and time. Regional variations of fatality-frequency distributions provide a measure of societal fragility. Self-similar behavior over many orders of magnitude in loss is indicative of a society that is much more resilient to natural disasters than one which exhibits a truncated fatality-frequency distribution. Upper truncated fatality-frequency distributions indicate the frequent occurrence of maximum fatality events and saturation in the degree of earthquake related damage for a region.

The second perspective compared changes in earthquake fatality-frequency distributions as a function of time for Italy, Japan, and China. Japan experienced a roll-back in the number and size of maximum fatality events in the twentieth century, while maintaining a relatively constant slope or D value. Italy and China, on the other hand, both experienced decreases in D during the twentieth century due to fewer small fatality events in Italy, and more large fatality events in China.

The third perspective examined natural disasters in the United States and identified two groups or families of disasters based on differences in scaling behavior. Earthquakes and hurricanes are typified by low frequency, high maximum fatality events; while floods and tornadoes are characterized as high frequency, low maximum fatality events.

In terms of future disasters, the self-similar scaling behavior of past losses can be used to anticipate and plan for those rare catastrophic events that incur the

TABLE 3 Poisson probability estimates for the occurrence of earthquake, hurricane, flood, and tornado disasters with fatality magnitudes of $F = 1$ and 3 (i.e., 10 and 1000 fatalities) in the United States during 1, 10, and 20 year exposure times, and estimates of the mean return period. Note the reversal in recurrence periods for $F = 1$ and $F = 3$ events. Floods and tornadoes have relatively shorter return periods for $F = 1$ events, whereas earthquakes and hurricanes have relatively short return times for $F = 3$ events.

Disaster	Exposure Time $F = 1$ (10 fatalities/event)			
	1 yr	10 yrs	20 yrs	Return Period, yrs
Earthquakes	0.11	0.67	0.89	9
Hurricanes	0.39	0.99	0.99	2
Floods	0.86	0.99	0.99	0.5
Tornadoes	0.96	0.99	0.99	0.3
Exposure Time Disaster	$F = 3$ (1000 fatalities/event)			
	1 yr	10 yrs	20 yrs	Return Period, yrs
Earthquakes	0.01	0.14	0.26	67
Hurricanes	0.06	0.46	0.71	16
Floods	0.004	0.04	0.08	250
Tornadoes	0.006	0.06	0.11	167

largest losses and place the most stress on society. Probabilistic estimates for the occurrence of catastrophic events in Table 2 provide a quantitative basis for prioritizing global earthquake disaster relief and mitigation programs. The probabilistic estimates in Table 3 provide the foundation for developing multidisaster mitigation programs at a national level. In both cases, these fatality-frequency distributions provide a framework for developing disaster mitigation strategies. Should disaster mitigation programs attempt to decrease the slope of the fatality-frequency curve by minimizing the number of small fatality events or increase the slope by trying to minimize the number and size of large fatality events? An alternate strategy would be to reduce the number of events in all fatality size classes, keeping the slope D constant.

ACKNOWLEDGMENTS

We thank R. Bilham, C. Bufe, J. Dewey, P. Dunbar, D. Mileti, D. Morton, D. Perkins, S. Tebbens, and L. Whiteside for data, stimulating conversations, and valuable reviews. Both S. P. N. and C. C. B. were supported by the United States Geological Survey G. K. Gilbert Fellowship Program for 1993–1994.

REFERENCES

1. Ambraseys, N. N., and C. P. Melville. *A History of Persian Earthquakes.* Cambridge: Cambridge University Press, 1982.
2. Barton, C. C., S. P. Nishenko, S. F. Tebbens, and W. A. Loeb. "Fractal Scaling and Forecasting of the Size and Frequency for Florida Hurricanes, 1886–1991 and of U.S. Hurricane Financial Loss, 1900–1989." This volume.
3. Bilham, R. "Earthquakes and Urban Growth." *Nature* **336** (1988): 625–626.
4. Coburn, A. W., A. Pomonis, and S. Sakai. "Assessing Strategies to Reduce Fatalities in Earthquakes." In *International Workshop on Earthquake Injury Epidemiology for Mitigation and Response*, edited by N. Jones, E. Noji, G. Smith and F. Krimgold, 107–132. Baltimore: John Hopkins University, 1989.
5. Dunbar, P. K., P. A. Lockridge, and L. S. Whiteside. *Catalog of Significant Earthquakes 2150 B.C. – A.D. 1991.* Boulder: U.S. Department of Commerce, 1992.
6. Grazulis, T. *Significant Tornadoes, 1680–1991.* St. Johnsbury, VT: Environmental Films, 1993.
7. Gutenberg, B., and C. F. Richter. "Frequency of Earthquakes in California." *Bull. Seismol. Soc. Amer.* **34** (1944): 185–188.

8. Hebert, P. L., and R. A. Case. "The Deadliest, Costliest, and Most Intense United States Hurricanes of This Century (and Other Frequently Requested Hurricane Facts)." NOAA Tech. Memorandum, NWS NHC 31. U.S. Department Commerce, Miami, 1990.

9. Ishimoto, M., and K. Iida. " Observations sur les Seismes Enregistres le Microsismographe Construit Dernierement." *Bull. Earth. Res. Inst., Tokyo Univ.* **17** (1939): 443–478.

10. Kagan, Y. Y., and L. Knopoff. "Spatial Distribution of Earthquakes: The Two-Point Correlation Function." *Geophys. J. R. Astr. Soc.* **62** (1980): 303–320

11. Knopoff, L., and Y. Y. Kagan. "Analysis of the Theory of Extremes as Applied to Earthquake Problems." *J. Geophys. Res.* **82** (1977): 5647–5657.

12. Mandelbrot, B. *The Fractal Geometry of Nature.* New York: W. H. Freeman, 1983.

13. McEvedy, C., and R. Jones. *Atlas of World Population History.* New York: Facts on File, 1979.

14. Pacheco, J. F., C. H. Scholz, and L. R. Sykes. "Changes in Frequency-Size Relationship from Small to Large Earthquakes." *Nature* **355** (1992): 71–73.

15. Richardson, L. F. *Statistics of Deadly Quarrels.* Pittsburgh, PA: Boxwood Press, 1960.

16. Richter, C. F. *Elementary Seismology.* New York: W. H. Freeman, 1958.

17. Smalley, R. F., Jr., J.-L. Chatelain, D. L. Turcotte, and P. Prevot. "A Fractal Approach to the Clustering of Earthquakes: Applications to the Seismicity of the New Hebrides." *Bull. Seism. Soc. Am.* **77** (1987): 1368–1381.

18. Snugg, A. L., and R. L. Carrodus. "Memorable Hurricanes of the United States since 1873." ESSA Tech. Memorandum WBTM SR-42, U.S. Department Commerce, Fort Worth, 1969.

19. Stover, C. W., and J. L. Coffman. *Seismicity of the United States 1568–1989 (Revised).* U.S. Geol. Survey Prof. Paper 1527. Washington, DC: U.S. Government Printing Office. 1993.

20. Turcotte, D. L. "A Fractal Approach to Probabilistic Seismic Hazard Assessment." *Tectonophysics* **167** (1989): 171–177.

21. Turcotte, D. L., and L. Greene,. "A Scale-Invariant Approach to Flood-Frequency Analysis." *Stochastic Hydrol. Hydraul.* **7** (1993): 33–40.

22. United Nations Department of International Economics and Social Affairs. *The Prospects of World Urbanization.* Population Studies No. 101, New York: United Nations, 1987

23. United States Army Corps of Engineers. *Annual Flood Damage Reports to Congress.* Washington, DC: U.S. Government Printing Office, 1986–1993.

Roger Bilham
CIRES and Department of Geological Sciences, University of Colorado, Boulder, CO 80309-0216

Global Fatalities from Earthquakes in the Past 2000 Years: Prognosis for the Next 30

By 2025 approximately 5500 million people will live in cities; more than the entire 1990 global rural and urban population. Dwelling units for this growing urban population will be constructed in the next 30 years in regions that are known to have experienced damaging historic earthquakes. In some locations the recurrence interval for damaging earthquakes is similar in time for the local population to have increased tenfold. In view of the association between earthquakes and the collapse of structures, this is cause for concern for it may be accompanied by a tenfold increase in fatalities from earthquakes. However, this forecast does not account for changes in urban construction methods that might reduce or intensify seismic risk, nor does it account for an unfavorable distribution of future earthquakes, that might aggravate seismic risk. This article reviews the population and earthquake fatality database for the past 2000 years to discern trends in global earthquake fatalities. The data appear to be reasonably complete only for the past 400 years, during which time the mean annual number of fatal earthquakes has increased roughly in proportion to population growth. The number of fatalities involved in these earthquakes, however, shows no simple relation to total population. For earthquakes in which fewer than 5000 fatalities occur, a smooth growth in their number can be used to forecast the probable future frequency of these events. However, earthquakes with fatalities exceeding 5000 contribute substantially to the historic fatality count. The cumulative fatality count for the past 400 years is apparently oblivious to the fourfold growth in global populations that has occurred during the present century. Given that this time is short compared to the recurrence interval of large earthquakes it is not unreasonable to conclude that the earthquake fatality rate will increase by an order of magnitude in the next century, should several large earthquakes occur near developing mega-cities.

GLOBAL URBANIZATION

Between B.C. 100 and about 1600 A.D. global populations doubled from perhaps 300 million to more than 600 million.[5] In the following 200 years, improvements in medicine and living conditions resulted in a dramatic reduction in mortality rates resulting in a second doubling in world population to 1200 million by 1800. Between 1800 and 1950 global populations doubled for the third time to 2500 million. The fourth doubling occurred in less than 40 years bringing global populations in 1990 to more than 5000 million. Although the rate of population increase has slowed, a fifth, and perhaps final, doubling of population in the next 50 years is projected.[11] While rural population growth continues, the bulk of world population growth is concentrated in urban areas in the developing nations. Urban populations in most developed nations are declining or showing minor growth (Los Angeles is an exception with a 1990 growth rate of 1.6%/year), but urban populations in the developing nations vary from 2% to 6%, driven both by internal growth and by a migration from the villages to the towns. In China, India, and Africa the rate of urban growth exceeds 4%, resulting in a doubling urban populations in less than 20 years.

Global urbanization is a new development for Homo Sapiens: prior to 1800 less than 2.5% of the world's population is estimated to have lived in cities. Diseases and epidemics moderated city populations and despite migration from surrounding rural communities, few cities exceeded 20,000.[9] Calder grimly observes that "until 1900 cities killed more people than they bred: they not only mopped up surplus agrarian population but buried much of it." The medicinal control of contagious diseases upset this balance, releasing uncontrolled urban growth. By 1930 the percentage of urban dwellers had increased sixfold, and by 1990, global cities contained 2400 million people (45% urban, 55% rural). By the year 2025, city population may exceed 5500 million (65% urban, 35% rural). The percentage of people living in urban communities in the developing nations is expected to increase from 37% in 1990 to 61% in 2025. Urban populations in the developed nations will increase from

73% (1990) to 83% in 2025.[11] These changing demographic trends are illustrated in Figure 2.

EARTHQUAKES

Indigenous materials are sometimes unsuited to the construction of single-story buildings resistant to horizontal shaking (e.g., adobe). Multistory dwellings may make excessive demands on otherwise tough building materials. Structural failure may be followed by fire, and sometimes by catastrophic flooding should a natural or artificial dam fail. However, prior to this century most fatal earthquakes typically buried people beneath their dwellings. The estimated cumulative number of fatalities attributable to earthquakes in the past 2000 years is close to 8 million with an estimated uncertainty of 11 million.[6] Dunbar et al.[6] include historical seismic data from many previous catalogs with the notable exception of Mallet's four volume catalog (1851–1854). A critical compilation to merge historical earthquake catalogs and eliminate duplicate and spurious entries is desirable but has not been attempted. In this article I have edited Dunbar et al.'s data principally by removing multiple entries. Most multiple entries offer several fatality counts of which the median value was usually selected, except where this appeared unreasonably high or low. In some cases multiple entries are not obvious by inspection, e.g., the 20 May 1202 Syrian event is manifest in six separate entries in some secondary sources; under three Muslim dates (B.C. 597–11 A.D.) and three Gregorian dates, 1202–1212.[1]

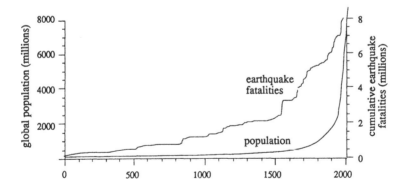

FIGURE 1   2000 years of population growth and earthquake fatalities. Data prior to 1600 must be treated with reservation, since both population estimates and fatalities from earthquakes are speculative. Data since 1600 are increasingly more complete.

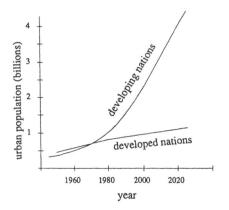

FIGURE 2   Past and projected urban populations in the developing and developed nations show radically different patterns. Despite the increase in urban populations in the developing nations the percentage of urban population in 2025 will represent only 65% of their total population, whereas the urban percentage in the developed nations already exceeds 73% (27% rural), and in 2025 it will have increased to 83% (17% rural).

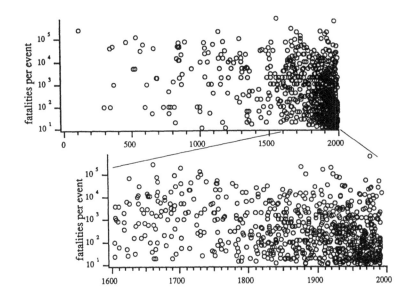

FIGURE 3   Global earthquake fatalities for the past 200 years, shown with an expanded scale for the past 400.

There may be other early events in the catalog that have similar multiplicity. The Calcutta 30 September 1737 earthquake was omitted,[4] as were all earthquakes that caused fewer than 10 fatalities. The edited list contains 1042 entries from 186 B.C. to A.D. 1994. Fatality estimates are few and unreliable before the seventeenth century, but improve in coverage following the Voyages of Exploration and the invention of printing. Nevertheless, even recent fatality counts may be unreliable, particularly for severe events, and the cumulative numbers may be in error by more than 20%.

A relation between the number of people at risk from earthquakes and the number of people on the planet is expected. This relation is manifest in Figures 1, 3, and 4. In Figure 1 an inflection in the rate of earthquake fatalities occurs in about 1500, before and after which the data approximate linear trends. This may correspond to the colonial expansions that occurred at this time resulting in more complete reporting. The European, Middle East, Japanese, and Chinese catalogs are reasonably complete for most of the past 2000 years, but India, the Americas, Southeast and Northeast Asia, south-Saharan Africa and Australasia are largely unrepresented prior to the sixteenth century. After 1500 the world became more widely explored and in 1600 numerical reports become available often in the form of administrative reports, and accounts of travelers.

Figure 3 shows that from 1500 to 1994 a tenfold increase in global populations from 0.7 billion to 7 billion was accompanied by a tenfold increase in the mean rate of fatal earthquakes from 0.6 per year prior 1500, to 6 per year at present. An understanding of trends in this period of time may thus be of value in interpreting the expected tenfold increase in populations for 1800–2100. Figure 3–5 also confirm that prior to 1500 there are insufficient events to provide meaningful statistics. That the number of fatal earthquakes should be proportional to total population is not entirely obvious since neither populations nor earthquakes are uniformly distributed on the earth's surface. Were earthquakes and people uniformly distributed on the Earth we should anticipate a relation between the total number of fatalities and world population. This relation is not evident in Figure 1. If anything the 5.7 million earthquake-related fatalities reported since 1500 show an approximately linear trend (11.4 k/yr) with large variance caused by the substantial contribution from infrequent disastrous events (Figure 5). For example, the mean global fatality rate for contiguous 125 year intervals since 1500 fluctuates with no simple relation to global population: 9 k/year 1500–1625, 13 k/year 1626–1750, 4 k/year 1751–1875, 20 k/year 1786–1995. Fatalities in the first and last interval are dominated by events in China where more than 0.5 million people died in single events.

The mean rate of earthquake fatalities since 1800 shows a clear increase in rate in Figure 5. However, the rate of increase is similar to one observed in 1650, when no quadrupling in global population occurred. In Figure 5(a), the increase in the number of fatalities per century is shown graphically for events with different degrees of mortality. The growth is steady for events with fewer than 10,000 fatalities, but this may also reflect reporting shortcomings for smaller events. The statistics for larger events might be considered insufficient to draw meaningful conclusions, yet the cumulative graph (Figure 5(b)) demonstrates that a smooth relation exists even for an epoch during which populations increase by a factor of 4. Nishenko and Barton[10] have examined similar fractal relations for different epochs as a function of region.

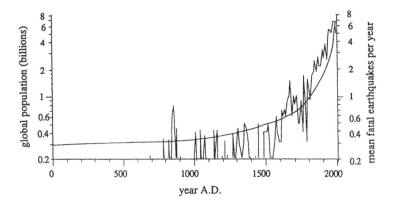

FIGURE 4   Global population compared to the mean number of fatal earthquakes per year. An order of magnitude increase in population in the past 400 years (600 million to 6000 million) is accompanied by a corresponding order of magnitude increase in fatal earthquakes (0.7/year to 7/year).

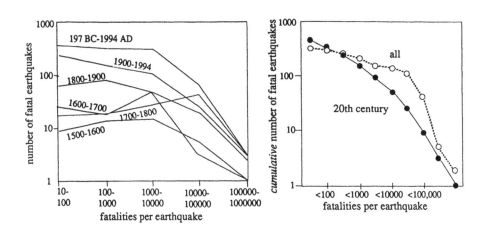

FIGURE 5   Relations between numbers of fatalities and number of earthquakes for different epochs 5(a), and the contribution to the cumulative number of fatalities from individual catastrophes 5(b).

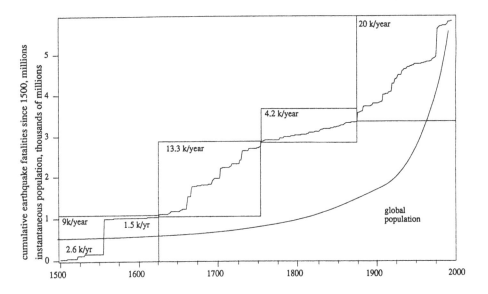

FIGURE 6   A closer view of cumulative fatalities in the past 500 years. The earthquake data are dominated by catastrophic earthquakes in China in the first and last 125-year intervals.

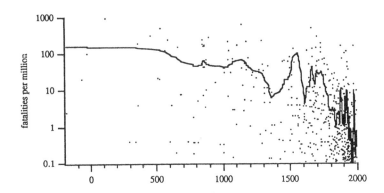

FIGURE 7   Fatalities from earthquakes normalized against global population. The solid line is a smoothed mean value. Each dot is an event in which the number of deaths has been divided by the inferred global population. The global mean fatality rate has fallen by approximately an order of magnitude since 1600 when records become reasonably reliable.

A curious result of increased global populations is that the fraction of the global population killed in events, has declined significantly in the past 200 years (Figure 7). Thus in one sense the risk to each individual on a global basis has declined. The data in Table 1 are interpolated using a smoothing spline to yield the estimated global population at the time of each earthquake, and the value plotted in Figure 7 is the quotient of the number of fatalities and the global population for each event. The mean value is obtained using a 50-point gaussian smoothing function. The units on the figure are earthquake fatalities per million people (fpm). The fpm falls from approximately 100 ftm (pre-1600) to 10–60 ftm in the following 200 years, to approximately 1 fpm at present.

TABLE 1 Global population figures used in this article. Pre-1950 (Calder[5]), post-1950 (United Nations[11]).

year	population (millions)
0	295
1000	325
1500	515
1750	770
1900	1680
1950	2516.44
1955	2752.11
1960	3019.65
1965	3336.32
1970	3697.85
1975	4079.02
1980	4448.04
1985	4851.43
1990	5292.19
1995	5770.29
2000	6260.8
2005	6739.23
2010	7204.34
2015	7659.86
2020	8091.63
2025	8504.22

DISCUSSION

Clearly, the fatality rate is coupled not so much to the mean global population but to the rate at which cities are shaken. The statistical fluctuations in the selective targeting of cities have similarities to the random targeting of nuclei in a particle accelerator. Borrowing from this analogy, if we assume that the diameter of the area of damaging shaking (say, Mercalli Intensity > VIII) represents the size of the "particle," and that each city is a target with a physical diameter of 1–50 km with a reaction "mass" proportional to population density, and separated from neighboring cities by distances very much larger than the size of each city, a reaction cross-section may be estimated to determine the probability of future interactions. The enlargement of urban centers increases the probability for future disaster, as, for example, the development of urban sprawl in Los Angeles that makes parts of this city vulnerable to moderate earthquakes simply because it has become spatially a larger target. The ill-advised development of Mexico City in a region sensitive to long-period shaking from distant events, makes urban agglomerations here a target for low-intensity shaking from several $M > 7.5$ events along the coast with a mean recurrence interval of approximately 30 years. However, the growth of cities from villages, which in the developing nations is often attended by the construction of multistory dwellings with little or no earthquake resistance, is probably a more important factor in increasing the probability of future fatal interactions between earthquakes and people. In some parts of the world, the recurrence of earthquakes which may have resulted in little damage to former villages may result in significant damage to the great cities that have now replaced them.

Although there is a random element in the targeting of large cities by large earthquakes, the cumulative fatalities involved in smaller events are sufficiently well behaved to form useful conclusions about future rates (Figure 8). For example, if all earthquakes with fatality counts exceeding 5000 people are removed from the catalog, a smooth increase in cumulative rate is obtained rising steadily from the sixteenth century to the present. Using this growth curve it is possible to estimate that in the next 30 years approximately 2000 ± 1300 people/year will be killed in this kind of disaster. The fatality rate in events involving fewer than 30,000 fatalities is approximately 5900 ± 1500 people/year. For larger events the curves become too irregular to provide a useful forecast.

Perhaps the most sinister societal development in terms of earthquake risk is the growth of supercities (populations exceeding 2 million), and megacities, which the UN define to contain more than 8 million people. In 1950 there were only two megacity agglomerations: London and New York. By 1990 there were 10, and in the year 2000 the UN has projected that 28 cities will attain this status (Figure 9). The Mexico City population will have reached 28.6 million people. The selective growth of many of the worlds largest cities near plate boundaries aggravates long-term urban seismic risk.[3]

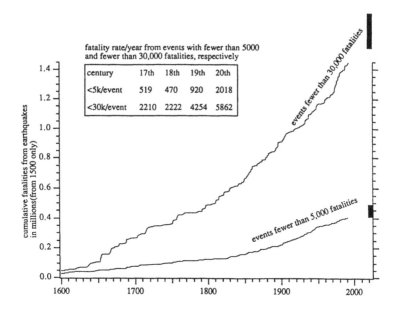

fatality rate/year from events with fewer than 5000 and fewer than 30,000 fatalities, respectively

century	17th	18th	19th	20th
<5k/event	519	470	920	2018
<30k/event	2210	2222	4254	5862

FIGURE 8 Earthquakes in which fatalities do not exceed 5000 and 30,000 respectively are sufficiently well behaved to forecast their probable rate early in the next century. For events involving fewer than 5000 fatalities the anticipated rate is 2020–1300/year; for events with fewer than 30,000 fatalities per year the rate is 5860–1500/year. For larger events the curves become irregular and consequently unpredictable.

The locations of disastrous earthquakes in the past 1000 years are shown in Figure 9 together with urban populations in cities greater than 1 million people projected for the year 2000. The total urban population held within these 353 cities is 1102 million. Approximately 30% of the world's megacities are located close to a strike-slip or convergent plate boundary where the largest earthquakes occur. Eighty percent of these are in developing nations housing approximately 80 million people. In cities of 2 million or more in seismogenic zones, 290 million people live. Many more live in cities with 1 million people.

Of particular concern are those cities that are also capital cities because fatalities and disruption in these cities are likely to jepordize the national economies of these countries. Mexico City and Teheran have populations that include substantial fractions of their national populations (24% and 12% respectively). Other countries may have an even greater fraction of their total population concentrated in one city: Santo Domingo (35%), Athens (37%), TelAviv (42%), Lima (31%), Santiago (36%).

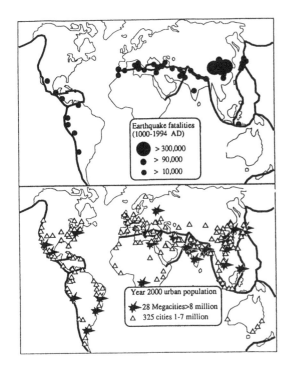

FIGURE 9 Fatalities from earthquakes in the past 1000 years compared to urban populations in cities greater than 1 million in the year 2000. More than half of the world's total population (6.3 billion) will live in cities: only the largest are shown here, housing 20% of the world's population.

CONCLUSIONS

The number of earthquakes resulting in fatalities has increased approximately in proportion to global populations, and although a decreasing fraction of the global population has been killed by earthquakes in this century compared to former centuries, seismic risk in certain regions has increased substantially. The cause of the apparent paradox lies in the growth of urban agglomerations where most of the world's growing population will live, and the location of many of these cities near plate boundaries where earthquakes occur quasi-periodically. The development of statistical predictions of seismic risk for these many urban areas must obviously be undertaken on a regional and local scale.

A few global trends are evident in the data of the past 500 years. The clearest is the number of fatalities involved in earthquakes resulting in fewer than 5000 fatalities. The cumulative number of deaths from these events has increased monotonically since 1600 (520 fatalities/year), to a late twentieth-century rate of 2020 fatalities per year. The rate is likely to increase moderately in the next 30 years. Similar growth is noted for earthquakes involving larger numbers of fatalities, although rates in future years are less easy to forecast because the curves become decreasingly regular. However, for earthquakes involving fewer than 30,000 fatalties the future rate is also fairly reliable rising by perhaps 30% from its current rate of approximately 6000 fatalities/year. Daunting though these figures are, these events do not contribute more than 30% to the past 125 years of earthquake fatalities $(= C520,000/\text{year})$.

The growth of giant urban agglomerations near regions of known seismic hazard is a new experiment for life on Earth. With few exceptions (Tangshan,[12] China, in 1976; Tokyo, Japan, in 1923), recent large earthquakes $(M > 7.5)$ have spared the world's major urban centers, this will not persist indefinitely. The recurrence interval for damaging earthquakes varies from 30 years to 3000 years and if population densities remain high in the next millennium, several megacities will be damaged by significant earthquakes. We are most certain of the fate of those cities near plate boundaries; however, mid-continent earthquakes also occur, albeit infrequently (c.f. $M > 8$ events at New Madrid and Charleston in the nineteenth-century U.S.), and these events will perhaps wreak great havoc in mid-continent cities where earthquake-resistant construction is not mandated. Until dwellings and civil structures are made resistant to earthquake shaking the earthquake fatality rate will remain high.

It is probable that the annual fatality rate from earthquakes will rise by a factor 4–10 in the next 30 years, attributable partly to an increase in the fatality rate from moderate earthquakes near large cities, but principally from a few catastrophic earthquakes near supercities (populations 2–28 million). Fatality counts exceeding 1 million from individual events are not unreasonable given that $> 50\%$ of an urban population can be lost in a single earthquake (as in Tangshan,[12] 1976) and that there are now dozens of cities near seismic belts with populations exceeding 2 million.

ACKNOWLEDGMENTS

I thank Susanna Gross for reading the manuscript. She, Lowell Whiteside, and Stuart Nishenko provided thoughtful insights concerning the interpretation of the data. NSF and the USGS supported the investigations.

REFERENCES

1. Ambraseys, N. N. "A Note on the Chronology of Willis's List of Earthquakes in Palestine and Syria." *Bull. Seism. Soc. Amer.* **52(1)** (1962): 77–80.
2. Ambraseys, N. N., and C. P. Melville. "An Analysis of the Eastern Mediterranean Earthquake of 20 May 1202." In *Proc. Symp. Historical Seismograms and Earthquakes, IASPEI/UNESCO Working Group on Historical Earthquakes,* edited by W. H. K. Lee, 390. (Tokyo Aug. 1985). San Francisco, 1987.
3. Bilham, R. "Earthquakes and Urban Development." *Nature* **336** (1988): 625–626.
4. Bilham, R. "The 1737 Calcutta Earthquake and Cyclone Evaluated." *Bull. Seism. Soc. Amer.* **84(5)** (1994): 1650–1657.
5. Calder, N. *Timescale,* 88. London: Chatto and Windus, 1984.
6. Dunbar, P. K., P. A. Lockridge, and L. S. Whiteside. *Catalog of Significant Earthquakes 2150 B.C. to A.D. 1991, Including Quantitative Casualties and Damage.* National Geophysical Data Center Report SE49, 320 Sept 1992.
7. Mallet, R. "Second Report on the Facts of Earthquake Phenomena." *Report of the Twenty-First meeting of the British Association for the Advancement of Science,* 272–320. (Ipswich, 1851), 272–320. 1852.
8. Mallet, R. "Catalogue of Recorded Earthquakes from 1606 B.C. to 1850 A.D. (continued from Report for 1853). *Report of the Twenty-Fourth meeting of the British Association for the Advancement of Science.* (Liverpool, 1854), 2–326. 1855.
9. McNeill, W. H. *Plagues and Peoples,* 340. 1976. Anchor Books, Reprint 1989.
10. Nishenko, S., and R. Buland. "Scaling Laws for Natural Disaster Fatalities." Santa Fe Institute Workshop, 1994.
11. United Nations. *World Urbanization Prospects 1990,* ST/ESA/SER.A/121., Sales No. E.91.XIII.11, pp. 223. New York: United Nations, 1991
12. Yong , C., K. Tsoi, C. Feibi, G. Zhenhuan, Z. Qijia, and C. Zhangli. *The Great Tangshan Earthquake of 1976—Anatomy of a Disaster.* Oxford: Pergamon Press, 1988.

Floods and Landslides

Leonard M. Sander
H.M. Randall Laboratory of Physics, The University of Michigan, Ann Arbor, MI 48109-1120

Diffusion-Limited Aggregation as a Paradigm for Modeling Dynamical Processes

We review the current state of knowledge about the diffusion-limited aggregation model which produces fractal patterns that are similar to the shapes of many naturally occurring objects. We point out the general features of the model which might carry over to models in earth sciences.

INTRODUCTION

The understanding of dynamical processes such as the growth of a crystal or of a river network are of central importance in many fields of pure and applied science. In recent years the availability of computer simulation has led many workers to devise simple models for such processes. In statistical physics the field of *pattern formation*[3,8,11,15,52] has had an explosion of activity mostly based on studies of models of this type, and in some fields of earth science such as geomorphology,[21,23,27,31,44,45,48,49] models with a similar spirit have had a good deal of success. The purpose of this paper is to review one of the best known statistical physics

Reduction & Predictability of Natural Disasters, Eds. Rundle, Turcotte, & Klein,
SFI Studies in the Sciences of Complexity, Vol. XXV, Addison-Wesley, 1996

models, diffusion-limited aggregation, and to point out which features of this model might give insight into processes that are important for the earth sciences and, eventually, for the understanding, say, of river networks or floods.

In the next section we will review some general features of kinetic models for growth and patterns from the point of view of a theoretical physicist. Then we will discuss the diffusion-limited aggregation (DLA) model as a case in point and show how it gives rise to disorderly fractal objects, and discuss some physical systems that seem to obey DLA rules. Next we will discuss some ways to arrive at a coarse-grained description of the aggregates formed. The last section is a summary.

KINETIC MODELS OF EXTENDED SYSTEMS

Modeling dynamics and pattern formation for an extended system far from equilibrium is a complicated subject. A major difficulty is that the simplicity of ordinary statistical mechanics and thermodynamics (which are only applicable in equilibrium) is lost. Let us recall that for an equilibrium system there is only one object that need be calculated, the partition function, or its logarithm, the free energy. The free energy obeys a minimum principle, and can be found by minimizing a candidate functional over all values of constraints. Near-equilibrium dynamics share some of this simplicity. For example, in the case where thermodynamic quantities, such as heat, flow from one place to another in a steady state, and the flows are *linear* in some external force then there is, once more, an extremal principle (for entropy production) that is obeyed.[42] However, the assumption of linearity is central to this construction.

It is important to realize that far from equilibrium all of this breaks down: as Cross and Hohenberg put it[8] "since we are not near equilibrium, there is no *a priori* reason to suppose that we have a Gibbs ensemble or a free-energy functional whose minima yield the patterns obtained under given external conditions." In general it is neither possible, nor desirable, to seek a minimum principle. Most workers in pattern formation study the kinetic rules themselves.

We will illustrate this with two examples taken from the theory of random, rough surfaces.[3,11] Consider a surface, $h(x, t)$, that is growing by the addition of matter and relaxes by local rearrangements: we could be considering sedimentation of sand in water where a grain drops and then rolls downhill if it can. A common way to model the evolution[10] is with a Langevin equation:

$$\frac{\partial h}{\partial t} = R + \eta + \nu \nabla^2 h,$$

$$= R + \eta + \frac{\delta F}{\delta h(x, t)} \, . \tag{1}$$

$$F = \int \left[\frac{\nu}{2}\right] |\nabla h|^2 dx$$

Here R is the mean growth rate, η is the noise due to the fluctuations in the growth, and the term in ν represents relaxation of the surface by rearrangement. In this case a generalized free-energy or Lyaponov functional, F, exists, and in the steady state (and in a moving frame where $h \to h - Rt$) $h(x)$ obeys an extremum principle except for fluctuations. Systems where this occurs are called gradient systems. (We should note that there are gradient systems for which the Langevin equation is nonlinear. That is not the case here.)

Now consider a slightly generalized model[22] which is appropriate for growth of amorphous solid films. This equation does *not* arise from variation of any F:

$$\frac{\partial h}{\partial t} = R + \eta + \nu \nabla^2 h + \lambda |\nabla h|^2 . \tag{2}$$

The extra term can be naturally generated in a variety of ways. For example, if we consider growth that is always normal to the local surface, it is easy to see that the growth rate in the upward direction is given by $R\sqrt{1 + |\nabla h|^2} \approx R + [R/2]|\nabla h|^2$. Thus the nonlinear term, $\lambda = R/2$, is growth-rate dependent, and vanishes in equilibrium, i.e., $R \to 0$. This model is believed to describe many real processes. Both Eqs. (1) and (2) give rise to self-affine fractal surfaces which are qualitatively similar. Nothing in the qualitative nature of the dynamics or the form of the surfaces produced changes because one system is a gradient system, and the other is not.

In pattern formation studies, most workers simply do not try to find a Lyaponov functional, and the theoretical techniques that are used are independent of its existence. It is almost certain that fractal DLA patterns do not correspond to the minimization of anything. However, certain river network models[21,44,45,48,49] call for the minimization of a nonlinear energy-dissipation function, and they seem quite successful in generating the observed fractal scaling and statistical properties of river basins. This seems very puzzling unless the success is an accident or illusory.

We can only speculate on the reasons for this success. It is possible that some argument could be made about separation of time scales, namely that networks develop sufficiently slowly compared to the water flow itself so that the far-from-equilibrium terms are absent. However, this seems to beg the question. We suggest that the most convenient way to analyze this question would be to attempt to invent a continuum description of river-basin development of the form of Eq. (1) that satisfies an optimization principle in the steady state, and then to see how robust the description is. This may prove to be difficult, however.

DIFFUSION-LIMITED GROWTH AND DLA

The DLA model is a simple idealization of a common natural process, the formation of natural objects where the rate-limiting step is often diffusion. In the simplest example (say, solidification from solution) particles random-walk and then stick to a growing aggregate. There are many other cases of this sort which are of interest: electrochemical deposition can in some cases be diffusion limited, and dielectric breakdown such as lightning is, in a generalized sense (as we will see below), an example of diffusion-limited growth. An example of the patterns formed is given in Figure 1.[43] It is remarkable how much this pattern resembles natural features of landscapes.

Diffusion-limited growth gives rise to remarkable morphologies which are often ramified, complex, and disorderly. This complexity is the major interest in the model. The origin of the complexity is in a *growth instability*: diffusion-limited growth is generically linearly unstable for flat growing surfaces. The patterns formed during growth are complicated and subtle because nonlinearities determine them completely. This sort of instability is probably a general feature of dynamical models which give rise to complex shapes, though, to our knowledge, such an analysis is not usually done in models for river basins.[21,23,27,31,44,45,48,49]

FIGURE 1 A diffusion-limited pattern of growth in the crystallization of GeSe2 from an amorphous film on a surface.[43] This deposit is polycrystalline and fractal.

We will be interested here in disorderly growth, but we should point out, for completeness, that another sort of pattern often arises, namely a dendritic shape, most commonly observed in a snowflake. In this case an overall anisotropy cures the disorder, and renders the patterns six-fold symmetric, but the complexity remains. If such anisotropy does not occur, diffusion-limited growth can give rise to fractals, which are our focus here.

THE STEFAN PROBLEM AND THE DIFFUSIVE GROWTH INSTABILITY

We will discuss here the classic linear stability analysis of the growth instability given by Mullins and Sekerka.[37] First we state the problem of diffusion-limited growth in continuum terms; this is known as the Stefan problem. Suppose the particles arriving at the aggregate have density $u(r, t)$. Then we have:

$$\frac{\partial u}{\partial t} = \nu \nabla^2 u, \tag{3}$$

$$v_n \propto -\nu \frac{\partial u}{\partial n}. \tag{4}$$

That is, u should obey the diffusion equation; ν is the diffusion constant. The normal growth velocity of the interface, v_n is proportional, on the average, to the flux onto the surface, $\partial u/\partial n$. It is useful to estimate the size of the term $\partial u/\partial t$ by noting that if there is a typical velocity of growth, V, then $\partial u/\partial t \sim V\partial u/\partial x$. Now:

$$|\nabla^2 u| \approx \left|\left(\frac{1}{l}\right) \frac{\partial u}{\partial x}\right| \tag{5}$$

where $l = \nu/V$, the diffusion length, sets the length scale for healing the diffusion field: outside the surface u returns to its asymptotic value as $e^{-r/l}$. In practical cases l is sometimes very much larger than the other scales in the problem, so that the right-hand member of Eq. (3) can be neglected. In this case Eq. (3) reduces to the Laplace equation, $\nabla^2 u = 0$ and u varies as a power law. We can imagine that we are to solve an electrostatics problem and advance the surface proportional to the electric field at each point. This is called the quasi-static regime.

The remaining equation necessary to describe the Stefan problem is a boundary condition on the field, u, at the interface of the growing object, and far away. The latter value gives the overall scale to the problem. The most usual form adopted for the boundary condition on the interface is: $u_s = d_0\kappa$ where d_0 is a length related to the surface tension, and κ is the curvature of the surface. In fact, it is necessary to have some cutoff, such as surface tension, in the problem. If we neglect surface tension, the electrostatic interpretation corresponds to the growth of a grounded conductor.

We ask, then, a simple question: why does Eq. (3) not simply describe a smooth surface which advances in time? In fact, if we start with a flat surface, it does advance in time (with $V \sim t^{-1/2}$) but this solution is not stable. The general method to handle situations of this sort is to put $h(x,t) = h_0(x - Vt) + \delta h(x,t)$ and find a linearized equation for δh. This is usually solved by putting $\delta h \propto e^{\omega t} e^{ikx}$ to find a "spectrum" $\omega(k)$. If $\omega > 0$ for some k, then we have a linear instability and the surface wrinkles on a scale $\lambda = 2\pi/k$. The steps in this procedure are easy to carry out, and are found in many references.[15,37]

It is sufficient for our purposes here to get the result with a simple argument. Suppose we start with a flat surface with a small bump of size λ and small height. The curvature is then small, so we can assume $u = 0$ on the surface. Consider the electrostatic interpretation above. We are asked to find the potential, u, near a grounded conductor with a bump, and then advance the different parts of the surface at a speed proportional to the surface electric field. Now the field is largest near the bump, this is the principle of the lightning rod. Thus the bump develops in time, faster than the rest of the surface, and becomes even sharper. In order to stabilize the growth, effects such as surface tension are necessary. And, clearly, these effects cannot stabilize large-scale instabilities. On some scale the surface must deform. As it deforms, the tip of the bump will, in general, split, and in the nonlinear regime (in the absence of anisotropy) the instabilities and splittings proliferate and mix. An important effect that comes into play is screening: adjacent tips will compete for incoming flux. The result of this complex dynamics is fractal growth, as we will now see.

FRACTAL GROWTH AND DLA

In the absence of anisotropy, and in the quasi-static regime, nothing controls the proliferation of tip splittings, and no steady state is possible. Instead a dynamic scaling structure arises which is very close to being a self-similar fractal. Many experimental patterns (Figure 1) have this property: the pattern in Figure 1 has a fractal dimension of approximately 1.7.

The best-known examples, such as fractal patterns in diffusion-limited growth, come from computer simulations of the diffusion-limited aggregation (DLA) model of Witten and Sander.[53,54] In this model, we launch a random walker and allow it to wander until it attaches to a growing cluster. Then, we launch another walker, and so on.

After a number of clusters are grown, it is quite easy to analyze the properties. For example, to estimate the fractal dimension, the simplest method is to measure the total mass, M, within a circle of radius R around the center of mass. Then[52,53,54] the fractal dimension, D, is given by fitting $M(R) \propto R^D$. One can also look for the typical power-law correlations of fractal geometry.[46]

DLA is a noisy version of the Stefan model. Equation (3) governs the probability density for finding a walker, though for the slow growth of the computer simulations we should set $\partial u/\partial t = 0$. The finite size of the walkers serves as a cutoff (albeit not of the form of surface tension). The boundary condition (4) is on the growth probability at a point on the cluster. The patterns seem to be self-similar fractals with fractal dimension of 1.7 in two dimensions and strongly resemble Figure 1: see Figure 2. This is a rather small cluster.

After more than a decade of work on the DLA model, we are still far from a complete understanding of this kind of fractal growth. The model is surprisingly difficult to deal with. The remaining questions about the model are of two types: (1) Are the observed numerically generated patterns really self-similar fractals? (2) Can we understand how the model gives fractals (without doing simulations)?

FIGURE 2   A small DLA cluster.

NUMERICAL STUDIES OF DLA. Though it is clear that the pattern of Figure 2 and larger ones of the same type are approximately self-similar, it is far from clear whether there are deviations from this simple behavior. For example, Mandelbrot and collaborators[30] have discovered a remarkable fact about very large two-dimensional DLA clusters by doing simulations up to 40,000,000 particles. They rescale and then coarse-grain images of the clusters (by surrounding every cluster point by a finite neighborhood) and find that the larger clusters seem to fill space. If this is correct, DLA clusters are not self-similar. In some sense the number of major branches grows with mass, and the cluster becomes more finely divided. This may be a crossover effect, but if it is asymptotically true, it is a severe challenge for theory.

There is a very long-standing question of just how homogeneous DLA clusters are. For example, some authors[34] claim that DLA is self-affine, and not self-similar. Hegger and Grassberger[20] have returned to this problem recently and found evidence that the previous results came from a slow crossover, and that very large clusters are locally self-similar.

The different branches in a cluster grow with markedly different rates because of the screening effect mentioned above. In fact, it appears that the growth probability is multifractal; the tips of the growth advance much, much faster than the fjords. The numerical proof of this statement was given by many authors.[1,18,35] The growth probability is found by either sending many random walkers to a fixed cluster and recording where they land, or directly solving the Laplace equation around a cluster, and finding the distribution of ∇u on the surface.

THEORY OF DLA. After more than a decade of frustration in formulating a theory for DLA, several groups now seem to have made substantial progress.[19,41,51] In our opinion, the most promising avenue is that of Halsey and co-workers who focus on the tip-splitting event as the fundamental building block of a theory. Based on this, Halsey gives a renormalization treatment of branching processes, which gives many of the observed features of DLA clusters and which could answer some of the long-standing questions in the field.

EXPERIMENTAL MANIFESTATIONS OF DLA

DLA-like patterns occur quite commonly in nature.[52] When do we expect to see such behavior? First, we must be in a situation with diffusion length that is long compared to the size of the deposit. Also, anisotropy must be either absent, or too small to overcome the averaging effects of extrinsic noise present in the experiment. If that is the case, tip splittings proliferate and produce fractals. For example, suppose the crystal is polycrystalline, with no long-range correlation for the crystalline axes. This is the case of the deposit of Figure 1: it is the formation of a polycrystalline deposit of $GeSe_2$ when an amorphous film rapidly crystallizes. In this case we observe well-defined DLA-like growths with fractal dimension ≈ 1.67.

ELECTROCHEMICAL DEPOSITION. Electrochemical deposition has been studied in detail in recent years in the context of DLA. Suppose we consider an experiment in which metal is deposited on a cathode from a solution, and the motion of the ions in solution is limited by diffusion. In this case we are dealing with an almost literal realization of the DLA model, provided the electric field in the solution is not important for the motion of the ions over most of the bulk. This is the case in the experiments of Brady and Ball[6] who considered the electrodeposition of copper from $CuSO_4$ and found fractal deposits.

Many experiments have been done using flat electrochemical cells[16,17,32,47] on which the driving force for the motion of the ions is the electric field. The ions follow field lines on large scales. We can reinterpret u to be the electrostatic potential which obeys the Laplace equation for a conductor taken to be an equipotential and note that the deposit grows when an ohmic current arrives at the surface. Thus we seem to be back in the same situation as before, and in some cases DLA-like deposits were observed. However, this experiment is very complicated and there are many parasitic effects which complicate the interpretation. However, in recent years several groups[12,13,14,24,25,36] have done very careful work on this problem and seem to have sorted out many of the remaining difficulties.

VISCOUS FINGERING AND DIELECTRIC BREAKDOWN. Viscous fingering is the unstable motion of the interface of two fluids, one more viscous than the other when the inviscid one is injected into the more viscous.[4] The fingering instability is a problem for oil-recovery processes in which air or water is injected into an oil field in order to drive the oil to a distant well; if the pattern of the interface is complex, it is difficult to know where to put the well. The phenomenon occurs for flow in a porous medium like an oil-bearing rock, and for flow between two parallel plates with a small gap, a so-called Hele-Shaw cell. The pattern of the relatively inviscid injected fluid plays the role of the crystal in the sections above, or the DLA cluster.

In order to relate this process to what we have discussed so far, we need only point out that flow in porous media or a Hele-Shaw cell is described by an empirical rule called D'Arcy's law which gives, for the fluid velocity, $V = -K\nabla u$ where u is the pressure in the viscous fluid. The proportionality constant depends on the viscosity. Since most fluids are almost incompressible, $\nabla \cdot V = -K\nabla^2 u = 0$. We take the zero of pressure to be that in the inviscid fluid (since it has small viscosity, its pressure there is approximately constant). We have $u_s = \sigma\kappa$ due to the pressure drop due to the curvature of the interface, where σ is the surface tension. At the interface D'Arcy's law reads $V_n = -K\partial u/\partial n$. These are identical to the Stefan problem in the quasi-static limit.[39,40] Experimental work in this area has been very extensive[5,9] and, for the radial viscous fingering case, patterns are produced that look very much like Figure 2 and have a fractal dimension that is very close to 1.7.

Another DLA-type process is the breakdown of a dielectric material when it is exposed to a very large electric field. Niemeyer et al.[38] showed that the pattern formed in some cases is exactly that of DLA. The connection to the model is once more through the continuum equations. Suppose we have a material with two electrodes with a large potential between them. Let u be the electrostatic potential between them. We have the Laplace equation for u. The breakdown channel where the material is highly ionized has a large conductivity, and thus a constant potential. And, it is reasonable to suppose (though may depend on the material in question) that the probability of breakdown is linear in the electric field, $-\nabla u$. Thus we have recovered Eqs. (3) and (4). Measured breakdown patterns, called Lichtenberg figures, have a fractal dimension of 1.7, and strongly resemble DLA.[38]

CONTINUUM THEORY OF DLA

As we noted above, it is most useful for general analysis if we can formulate a theory in terms of a continuum theory, because then we can see large-scale features most easily. A beginning in this direction is to inquire whether there is a mean-field theory of some sort for the growth. There is, indeed, such a theory, which was suggested by Witten and Sander[54] and developed by several groups.[2,54]

In this approach we coarse-grain the density of the aggregate, and average over angles. Thus we seek the average profile of a cluster. The density is replaced by a function $\rho(r,t)$, which is coupled to the random-walker probability function, u, which we have discussed up to now. Since the walker can be absorbed inside the coarse-grained profile, we treat the change in the density as a sink for particles. Thus we have instead of the Laplace equation with a boundary condition:

$$\nabla^2 u = \frac{\partial \rho}{\partial t} . \tag{6}$$

We now must describe the probability of absorption. Since a particle is absorbed only where there is matter, we are tempted to set this to be $u\rho$. However, a bit of thought will show that this will not do: the finite particle size, a, is missing here: we must allow the absorption to allow growth into a region where ρ is initially zero. The simplest way to represent this is to write:

$$\frac{\partial \rho}{\partial t} = u[\rho + a^2 \nabla^2 \rho] . \tag{7}$$

These two equations are to be solved for the profile. Well inside the aggregate, $u \to 0$ exponentially fast since the random walkers are being absorbed by the cluster. In fact, since $\nabla^2 u/u \approx \rho$, we can define a screening length $\xi \sim 1/\sqrt{\rho}$. Outside we assume $u \to u_o$.

Now the approximate solution of Ball, Nauenberg, and Witten[2,54] is based on the observation that the solution to Eq. (7) in the exterior region (where we can take $u \approx u_o$) is, up to power laws in t:

$$\rho \sim \exp\left(t - \frac{r^2}{4a^2 t}\right) . \tag{8}$$

This is to be matched to the interior solution in the surface region, i.e., at some average radius, R. A reasonable definition of R is the point where the profile becomes steep enough so that it changes within a screening length, ξ, so that we can no longer solve Eq. (7) with constant u. Thus:

$$\frac{[1}{\rho]}\frac{\partial\rho(R)}{\partial r} \sim \frac{1}{\xi} \sim \sqrt{\rho}. \tag{9}$$

Combining these two equations gives for the mean radius, R,

$$\frac{R^2}{4a^2t^2} \sim \exp\left(t - \frac{R^2}{4a^2t}\right). \tag{10}$$

Now this matching can only be consistent if $R = 2at$. Thus, in this model the velocity is constant. Remarkably, this fact is enough to give the fractal dimension.

This is easy to see: Consider a large sphere of radius r which is far away from the cluster. The solution to Eq. (6) near the outer sphere (where $\rho \approx 0$) is $u \approx u_0[1 - (R/r)^{d-2}]$. (In two dimensions the reasoning must be altered slightly, but the final result is the same.) The total amount of matter added, dM/dt, is the integral over the large sphere of ∇u; i.e., $dM/dt \sim R^{d-2}$. Now $R \propto t$. Thus $M \propto R^{d-1}$, and the fractal dimension is $d - 1$. It seems that for large d this is a reasonable estimate.[33]

A very recent series of papers[7,28] seem to show that we can go beyond mean-field theory by turning Eq. (7) into a Langevin equation which for a channel geometry takes the form:

$$\partial\rho/\partial t = u[\rho^\gamma + a^2\nabla^2\rho] + \eta. \tag{11}$$

Here γ is a number near unity, and η is a random noise. It will be extremely interesting to see how this subject develops, because we expect to make a sensible comparison of theories that generate fractals only at a coarse-grained level.

SUMMARY

In this chapter we have tried to give a brief review of some of the work on DLA with the hope that it can serve as a case study for developing models to describe more complex situations that are found in earth sciences. DLA is a good candidate to be such an example because it is remarkably simple, physically motivated, and yet naturally generates fractals very like natural fractal aggregates. It is well known that fractal properties are very relevant in geology[50] and understanding of these how these properties arise could certainly help in projects that are relevant to the subject of this volume, controlling natural disasters.

It is very hard to predict which, if any, of the developments that we have described will be relevant to models of this sort. Still, we would venture to guess that three of the ideas presented above will prove relevant. They are: (1) analysis of growth instabilities, (2) formulation of a continuum theory, and (3) description of the applications of the model in terms of the fundamental physical processes.

1. Instabilities of development of a simple shape are probably an ingredient in any model that gives rise to complexity. We have[29] given a series of such analyses for several models, and gleaned a good deal of insight in this way. See also Godreche[15] for more applications.

2. Continuum equations have, in almost every case, given more tractable models than discrete ones, though the discrete models are more useful for computer simulation. For example, though we have not discussed this, an essentially complete treatment of the roughening of thin films has been possible by studying equations such as Eq. (2): see Family and Vicsek[11] and Barabasi and Stanley.[3] We have devised a method[26] for deriving a continuum theory by examining discrete data. This subject is still under development.

3. In devising a model to describe a given physical situation, there is usually no substitute for sticking to the actual processes. The whole art of modeling, of course, is to discard all but the *relevant* details. Of course, figuring out what is relevant is just what makes the whole subject interesting, and difficult.

ACKNOWLEDGMENT

This work is supported by Department of Energy grant DE-FG02-95ER45546. I would like to thank Paul Meakin, Fred Mackintosh, and many participants in the workshop that gave rise to this volume for helpful discussions.

REFERENCES

1. Amitrano, C., A. Coniglio, and F. diLiberto. "Growth Probability Distribution in Kinetic Aggregation Processes." *Phys. Rev. Lett.* **57** (1987): 1016.
2. Ball, R., M. Nauenberg, and T. Witten. "Diffusion-Controlled Aggregation in the Coninuum Approximation." *Phys. Rev. A* **29** (1984): 2017.
3. Barabasi, A., and H. E. Stanley. *Fractal Concepts in Surface Growth.* Cambridge: Cambridge University Press, 1995
4. Bensimon, D., L. Kadanoff, S. Liang, B. Shraiman, B., and C. Tang. "Viscous Flows in 2 Dimensions." *Rev. Mod. Phys.* **58** (1986): 977.

Anissotropy in Interfacial Pattern Formation." *Phys. Rev. Lett.* **55** (1985): 1315.

6. Brady, R., and R. C. Ball. "Fractal Growth of Copper Electrodeposits." *Nature (London)* **309** (1984): 225.

7. Brener, E., H. Levine, and Y. Tu. "Mean-Field Theory for Diffusion-Limited Aggregation in Low Dimensions." *Phys. Rev. Lett.* **66** (1991): 1978.

8. Cross, M., and P. Hohenberg. "Pattern Formation out of Equilibrium." *Rev. Mod. Phys.* **65** (1993): 851.

9. Couder, Y. "Viscous Fingering." In *Random Fluctuations and Pattern Growth*, edited by H. E. Stanley and N. Ostrowsky. Boston: Kluwer, 1988.

10. Edwards, S. F., and D. R. Wilkinson. "The Surface Statistics of a Granular Aggregate." *Proc. Roy. Soc. Lond.* **A381** (1982): 17.

11. Family, F., and T. Vicsek. *Dynamics of Fractal Surfaces.* Singapore: World Scientific, 1991.

12. Fleury, V., J.-N. Chazalviel, M. Rosso, and B. Sapoval. "The Role of the Anions in the Growth Speed of Fractals Electrodeposits." *J. Electroanal. Chem.* **290** (1990): 249.

13. Fleury, V., M. Rosso, and J.-N. Chazalviel. "Geometrical Aspect of Electrodeposition: The Hecker Effect." *Phys. Rev. B* **43** (1991): 690.

14. Garik, P., D. Barkley, E. Ben-Jacob, E. Bochner, N. Broxholm, B. Miller, B. Orr, and R. Zamir. "Laplace and Diffusion-Field Controlled Growth in Electrochemical Depossition." *Phys. Rev. Lett.* **62** (1989): 2703.

15. Godreche, C., ed. *Solids far from Equilibrium.* Cambridge: Cambridge University Press 1992.

16. Grier, D., E. Ben-Jacob, R. Clarke, and L. M. Sander. "Morphology and Microstructure in Electrochemical Deposition of Zinc." *Phys. Rev. Lett.* **56** (1986): 1264.

17. Grier, D. G., D. A. Kessler, and L. M. Sander. "Stability of the Dense Radial Morphology in Diffusive Pattern Formation." *Phys. Rev. Lett.* **59** (1987): 2315.

18. Halsey, T., P. Meakin, and I. Procaccia. "Scaling Structure of the Surface Layer of Diffusion-Limited Aggregates." *Phys. Rev. Lett.* **56** (1986): 854.

19. Halsey, T. "Diffusion-Limited Aggregation as Branched Growth." *Phys. Rev. Lett.* **72** (1994): 1228.

20. Heeger, R., and P. Grassberger. "Is Diffusion-Limited Aggregation Locally Isotropic or Self-Affine?" *Phys. Rev. Lett.* **73** (1994): 1672.

21. Howard, A. "Theoretical Model of Optimal Drainage Networks." *Water Resour. Res.* **7** (1990): 863.

22. Kardar, M., G. Parisi, and Y.-C. Zhang. "Dynamic Scaling of Growing Interfaces." *Phys. Rev. Lett.* **56** (1986): 889.

23. Kramer, S., and M. Marder. "Evolution of River Networks." *Phys. Rev. Lett.* **68** (1992): 205.

24. Kuhn, A., and F. Argoul. "Influence of Chemical Perturbations on the Surface Roughness." *Fractals* **1** (1993): 451.

25. Kuhn, A., and F. Argoul. "Spatiotemporal Morphological Transitions in Thin-Layer Electrodeposition." *Phys. Rev. E* **49** (1994): 4298.
26. Lam, C-H., and L. M. Sander. "Inverse Method for Interface Problems." *Phys. Rev. Lett.* **71** (1993): 561.
27. Leheny, R., and S. Nagel. "Model for the Evolution of River Networks." *Phys. Rev. Lett.* **71** (1993): 1470.
28. Levine, H., and Y. Tu. "Theory of Diffusion-Limited Growth." *Phys. Rev. E* **48** (1993): R4207.
29. Louis, E., O. Pla, L. M. Sander, and F. Guinea. "Variations on the Theme of Diffusion-Limited Growth." *Mod. Phys. Lett. B* **8** (1994): 1739.
30. Mandelbrot, B. "Plane DLA is not Self-Similar; Is It a Fractal That Become Increasingly Compact as it Grows." *Physica A* **191** (1992): 95.
31. Masek, J., and D. Turcotte. "A Diffusion-Limited Aggregation Model for the Evolution of Drainage Networks." *Earth and Planet. Sci. Lett.* **119** (1993): 379.
32. Matsushita, M., M. Sano, Y. Hayakawa, H. Honjo, and Y. Sawada. "Fractal Structures of Zinc Metal Leaves Grown by Electrodeposition." *Phys. Rev. Lett.* **52** (1984): 286.
33. Meakin, P. "Diffusion-Controlled Cluster Formation in 2-6 Dimensional Space." *Phys. Rev.* **A27** (1983): 604.
34. Meakin, P., and T. Vicsek. "Internal Structure of Diffusion-Limited Aggregates." *Phys. Rev.* **A32** (1985): 685.
35. Meakin, P., A. Coniglio, H. E. Stanley, and T. Witten. "Scaling Properties for the Surfces of Fractal and Nonfractal Objects." *Phys. Rev. A* **34** (1986): 3325.
36. Melrose, J. R., D. B. Hibbert, and R. C. Ball. "Interfacial Velocity in Electrochemical Deposition and the Hecker Transition." *Phys. Rev. Lett.* **65** (1990): 3009.
37. Mullins, W. W., and R. F. Sekerka. "Stability of a Planer Interface During Solidification of a Dilute Alloy." *J. Appl. Phys.* **34** (1963): 323.
38. Niemeyer, L., L. Pietronero, and H. J. Wiesmann. "Fractal Dimension of Dielectric Breakdown." *Phys. Rev. Lett.* **52** (1984): 1033.
39. Paterson, L. "Radial Fingering in a Hele-Shaw Cell." *J. Fluid Mech.* **113** (1981): 513.
40. Paterson, L. "Diffusion-Limited Aggregation and Two-Field Displacement sin Porous Media." *Phys. Rev. Lett.* **52** (1984): 1621.
41. Pietronero, L., A. Erzan, and C. Evertsz. "Theory of Fractal Growth." *Phys. Rev. Lett.* **61** (1988): 861.
42. Prigogine, I. *From Being to Becoming*, 88. San Francisco, CA: W. H. Freeman, 1980.
43. Radnoczi, G., T. Vicsek, L. Sander, and D. Grier. "Growth of Fractal Crystals in Amorphous GeSe$_2$ Films." *Phys. Rev. A* **35** (1987): 4012.

44. Rinaldo, A., I. Rodriguez-Iturbe, R. Rigon, R. L. Bras, E. Ijjasz-Vasquez, and A. Marani. "Minimum Energy and Fractal Structures of Drainage Networks." *Water Resour. Res.* **28** (1992): 1095.

45. Rinaldo, A., I. Rodriguez-Iturbe, R. Rigon, E. Ijjasz-Vasquez, and R. L. Bras. "Self-Organized Fractal River Networks." *Phys. Rev. Lett.* **70** (1993): 822.

46. Sander, E., L. M. Sander, and R. M. Ziff. "Fractals and Fractal Correlations." *Comp. Phys.* **8** (1994): 420.

47. Sawada, Y., A. Dougherty, and J. Gollub. "Dendritic and Fractal Patterns in Electrolytic Metal Deposits." *Phys. Rev. Lett.* **56** (1986): 1260.

48. Sun, T., P. Meakin, and T. Jossang. "Minimum Energy Dissipation Model for River Basin Geometry." *Phys. Rev. E* **49** (1994): 4865.

49. Troutman, B., and M. Karlinger. "Gibbs Distribution on Drainage Networks." *Water Resour. Res.* **28** (1992): 563.

50. Turcotte, D. *Fractals and Chaos in Geology and Geophysics.* Cambridge, Cambridge University Press, 1992.

51. Vespignani, A., and L. Pietronero. "Fixed Scale Transformation Applied to Diffusion-Limited Aggregation and Dielectric Breakdown Model in Three Dimensions." *Physica A* **173** (1991): 1.

52. Vicsek, T. ' *Fractal Growth Phenomena.* Singapore: World Scientific, 1992.

53. Witten, T. A., and L. M. Sander. "Diffusion-Limited Aggretation, A Kinetic Critical Phenomenon." *Phys. Rev. Lett.* **47** (1981): 1400.

54. Witten, T. A., and L. M. Sander. "Diffusion-Limited Aggregation." *Phys. Rev. B* **27** (1983): 2586.

D. L. Turcotte and K. Haselton
Department of Geological Sciences, Cornell University, Ithaca, NY 14853

A Dynamical Systems Approach to Flood-Frequency Forecasting

Floods and droughts constitute extreme events of great consequence to society. A wide variety of statistical techniques have been applied to the evaluation of the flood hazard but with questionable success. A primary difficulty is the relatively short time span over which historical data is available, and quantitative estimates for paleofloods are generally suspect. It was in the context of floods that Hurst introduced the concept of the rescaled range. This was subsequently extended by Mandelbrot and his colleagues to concepts of fractional Gaussian noises and fractional Brownian walks. These studies introduced the controversial possibility that the extremes of floods and droughts could be fractal. In order to study historical flood-frequency records, we plot the log of the number of floods on a river per unit time in which the peak discharge exceeds a specified value against the log of that value. For ten benchmark stations we find good correlations with scale-invariant (fractal) statistics. We suggest that the underlying physical processes associated with the generation of floods are sufficiently scale invariant over time scales from one to one hundred years that they provide a rational basis for the application of scale-invariant statistics. We propose that the ratio of the ten-year peak discharge to the one-year peak discharge F is also a quantitative measure of flood potential. With scale invariance, F is also the ratio of the one-hundred-year flood to

Reduction & Predictability of Natural Disasters, Eds. Rundle, Turcotte, & Klein,
SFI Studies in the Sciences of Complexity, Vol. XXV, Addison-Wesley, 1996   **51**

the ten-year flood. We find that the values of F for ten stations on rivers throughout the country range from 2.04 to 8.11 and find strong regional variations that can be correlated in terms of climate. We have also carried out R/S analyses for the ten stations and have obtained values of the Hurst exponent H_1, the values of H_1 for the ten stations range from 0.66 to 0.73 indicating moderate persistence. Flood frequencies are sensitive to the skewness of the basic distribution, i.e., F (the Noah effect), and are not sensitive to persistence, i.e, H_1, (the Joseph effect). In order to test the validity of the fractal flood frequency hypothesis, we consider the fractal fits to 41-year records for 1009 USGS streamflow stations that are unaffected by flood control projects; good fits are obtained. Consistent regional variations in the flood intensity factor are also found.

INTRODUCTION

The flow in a river is a classic example of a time series and the peaks in this time series constitute floods. Floods are a major hazard to many cities and estimates of the flood hazard have major economic implications for the location of buildings, the construction of flood control levees, and the provision of flood insurance. The time integral of the flow in a river is required for the design of reservoirs and to assess available water supplies during periods of drought.

Quantitatively the definition of a flood is not a trivial problem. The river flow time series $\dot{V}(t)$ will generally have a strong annual component; in the United States this is taken account of through the water year which begins on October 1. The peak discharge at a measuring station during a water year $\dot{V}_{am}$ is taken to be the annual flood by definition. However, the use of the annual flood for statistical considerations strongly biases the results. In some years the annual flood will be much smaller than a number of "statistically" independent floods in other years. But the key question is how to define statistically independent floods. The method used in this paper is to eliminate a period of two months on either side of a maximum annual flood and then look for a secondary flood, and so forth. A second approach would be to require a drop in flow to some specified fraction of the peak flow, say 50%.

In order to estimate the severity of future floods, historical records are used to provide flood-frequency estimates. Unfortunately, this record generally covers a relatively short time span and no general basis has been accepted for its extrapolation. Quantitative estimates of peak discharges associated with paleofloods are generally not sufficiently accurate to be of much value.

A wide variety of geostatistical distributions have been applied to flood-frequency forecasts, often with quite divergent predictions. Examples of distributions used include power law (fractal), lognormal, gamma, Gumbel, log-Gumbel, Hazen, and log-Pearson type III. A comparison of these distributions (except the power law) was given by Bensen[1] for ten benchmark streamflow stations. This comparison was extendedfloods, forecasting to include the power-law distribution by Turcotte and Greene.[12] On the basis of such comparisons the log-Pearson type III was adopted as the official U.S. basis for flood-frequency forecasting.[13] Previous work on this problem has been treated on a strictly empirical basis. The available data have been fit with the statistical distributions using various moments of the observations to constrain the available constants. Also, conditions have been specified for the elimination of outliers.

A basic question is whether there is a physical basis for analyzing the flood-frequency problem, and in particular, whether fractal statistics would be expected to be applicable. Indications that such an approach might be successful comes from the work of Hurst.[4,5,6] Hurst spent his life studying the flow characteristics of the Nile and introduced the rescaled range (R/S) analysis. He found that the variations in the reservoir storage (the range) scaled with the time interval considered as a power law. Mandelbrot and Wallis[7,8,9,10] introduced the concepts of fractional Gaussian noises and fractional Brownian walks; both are recognized as fractal distributions. The fractional Brownian walks are the integrals of the fractional Gaussian noises and yield a power-law relation between R/S and T. Mandelbrot and Wallis[10] considered noises with non-Gaussian distributions and obtained similar results. They also introduced the Noah and Joseph effects. The Noah effect is the skewness of the distribution of flows in a river (or of a non-Gaussian distribution) and the Joseph effect is the persistence of the flows. Although the concepts introduced by Hurst and Mandelbrot and Wallis have been considered in a wide variety of applications, they have not influenced approaches to flood-frequency forecasting. This point will be a central feature of this paper along with a general discussion of the applicability of fractal statistics.

ANALYSIS

The volumetric flow in a river as a function of time is given by $\dot{V}(t)$. We define floods to be the maximum values of $\dot{V}(t)$, $\dot{V}_m$, which are separated from each other by at least two months. The definition of a fractal time series from the correlations between the flows at times t and at times $t + T$ is given by

$$\Pr\left[\frac{\dot{V}(t+T) - \dot{V}(t)}{T^H} \leq y\right] = f(y) \tag{1}$$

where $f(y)$ is a normalized cumulative probability distribution which is often a Gaussian but can be a lognormal or other distribution. The scaling constant H is known as the Hausdorff measure and can be related to the fractal dimension D by $D = 2 - H$[11].

If the nonstationary time series satisfies Eq. (1), then its extreme value $\dot{V}_m$ in a time interval T is related to the interval by

$$\dot{V}_m(T) = C_1 T^H. \tag{2}$$

Since river discharges have a strong annual variability, the interval T is generally taken as an integer number of years when floods are considered. This scale-invariant distribution can also be expressed in terms of the ratio F of the peak discharge over a ten-year period to the peak discharge over a one-year period. With self-similarity the parameter F is also the ratio of the 100-year peak discharge to the 10-year peak discharge and the ratio of the 1000-year peak discharge to the 100-year peak discharge. In terms of H and D we have

$$F = 10^H = 10^{2-D}. \tag{3}$$

Combining Eqs. (2) and (3) and introducing the ten-year flood as a reference flood gives

$$\dot{V}_m(T) = \dot{V}_m(10) \left(\frac{T}{10}\right)^{\log_{10} F}. \tag{4}$$

This relation will be correlated with flood data and the values of F will be obtained.

Before considering actual examples we will also introduce rescaled range (R/S) analysis. Hurst[4,5,6] proposed this empirical approach to the statistics of floods and draughts. The method is illustrated in Figure 1. Consider a reservoir behind a dam that never overflows or empties, the flow into the reservoir is $\dot{V}(t)$ and the flow out of the reservoir is $\overline{V}(T)$ defined by

$$\overline{V} = \frac{1}{T} \int_0^T \dot{V}(t) dt. \tag{5}$$

The volume of water in the reservoir $V(t)$ is given by

$$V(t) = V(0) + \int_0^t \dot{V}(t') dt' - t\overline{V}(T) \tag{6}$$

and the range is defined by

$$R(T) = V_{\max} - V_{\min} \tag{7}$$

where $V_{\max}$ is the maximum and $V_{\min}$ the minimum volume stored during the interval T. The rescaled range is defined as R/S where S is the standard deviation of the flow during the period T

$$S(T) = \left[\frac{1}{T} \int_0^T \left(\dot{V}(t) - \overline{V}\right)^2 dt\right]^{1/2}. \tag{8}$$

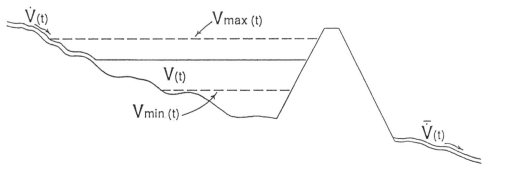

FIGURE 1 Illustration of how rescaled range (R/S) analysis is carried out. The flow into a reservoir is $\dot{V}(t)$ and the flow out is $\bar{V}(T)$. The maximum volume of water in the reservoir during the period T is $V_{\max}(T)$ and the minimum $V_{\min}(T)$; the difference is the range $R(T) = V_{\max}(T) - V_{\min}(T)$.

Hurst[4,5,6] found that for many time series the rescaled range satisfies the empirical relation

$$\frac{R}{S} = \left(\frac{T}{2}\right)^{H_1} \tag{9}$$

where H_1 is known as the Hurst exponent. Examples included river discharges, rainfall, varves, temperatures, sunspot numbers, and tree rings. In many cases the value of the Hurst exponent is near 0.7.

If a Gaussian white noise sequence of numbers is summed, the result is a Brownian walk. An R/S analysis of the white noise sequence gives a Hurst exponent $H_1 = 0.5$; thus the Hurst exponent is equal to the Hausdorff measure of the integrated signal, a Brownian walk with $H = 0.5$. Mandelbrot and Wallis[7,8,9,10] introduced the concept of fractional Gaussian noises and their integrals, fractional Brownian walks. They showed that the Hurst exponent H_1 of a fractional Gaussian noise is equal to the Hausdorff measure of the corresponding fractional Brownian walk.

If $0.5 < H_1 < 1$, the original time series is said to be persistent; adjacent values are more strongly correlated than if they were random. The higher the value of H_1, the greater the persistence. If $0 < H_1 < 0.5$, the original time series is said to be antipersistent; adjacent values are less correlated than if they were random.

RESULTS

We now turn to the analysis of flood-frequency records. As our first example, the ten benchmark stations considered by Benson[1] will be studied. Benson[1] applied a variety of geostatistical distributions to the data from these stations; these will be compared with the fractal approach discussed above. The maximum annual floods for two stations are given in Figure 2. For our analysis we consider floods that are separated by more than two months as discussed above.

The largest floods for each record are ordered, the largest flood is assigned a period equal to the length of the record τ_0, the second largest flood a period to $\tau_0/2$, the third largest flood a period $\tau_0/3$, and so forth. The log of the peak discharge for each flood is plotted against the log of its assigned period. Results for station 1-1805 on the Middle Branch of the Westfield River in Goss Heights, Massachusetts are given in Figure 3(a) for the period 1911–1960,[3] the solid line is the least-square fit of Eq. (2) with the data over the range $50 < V < 200$ m^3/s; large floods are omitted from the fit because of their small number. The solid line corresponds to $H = 0.51$ and from Eq. (3) we have $F = 3.3$. Results for station 11-0980 in the Arroyo Seco near Pasadena, California are given in Figure 3(b) for the period 1914–1965,[15] the solid line is the best fit of Eq. (2) with the data over the range $10 < V < 100$ m^3/s. The solid line corresponds to $H = 0.87$ and from Eq. (3) we have $F = 7.4$. In both cases the fit to the scale-invariant (fractal) relation is quite good. The values of H and F in California are considerably larger than in Massachusetts. Large floods are relatively more probably in the arid climate than in the temperate climate.

The values of H, D, and F are given for all ten benchmark stations in Table 1. The correlations with the fractal relation (2) in Figure 3 are typical of the ten stations. The parameter F is a measure of the relative severity of flooding. The higher the value of F, the more likely that severe floods will occur. Our results show that there are clear regional trends in values of F. The values in the southwest including Nevada ($F = 4.13$) and New Mexico ($F = 4.27$) as well as California ($F = 7.4$) are systematically high. The high values can be attributed to the arid conditions and the rare tropical (monsoonal) storm that causes severe flooding. Central Texas ($F = 5.24$) is also high and Georgia ($F = 3.47$) is intermediate. These areas are influenced by hurricanes. The northern tier of states including Massachusetts ($F = 3.26$), Minnesota ($F = 2.95$), Nebraska ($F = 3.47$), and Wyoming ($F = 3.31$) range from low values in the east to intermediate values in the west. Washington ($F = 2.04$) has the lowest value of the stations considered; this low value is consistent with the maritime climate where extremes of climate are rare.

We have also determined the Hurst exponent for the ten benchmark stations. Values of R/S for $T = 5$, 10, 25, and 50 years ($R/S = 1$ for $T = 2$ by definition) are given in Figure 4(a) for station 1-1805 (Westfield, MA) and in Figure 4(b) for station 11-0980 (Pasadena, CA). Good correlations are obtained with Eq. (9) taking $H_1 = 0.67$ for station 1-1805 and $H_1 = 0.68$ for station 11-0980. Values of H_1 for all ten stations are given in Table 1. The values are nearly constant with a range from

0.66 to 0.73 indicating moderate persistence. It is not surprising that the values of the Hausdorff measure H differ from the values of the Hurst exponent H_1 since the former refers to the statistics of the flood events and the latter to the statistics of the running sum.

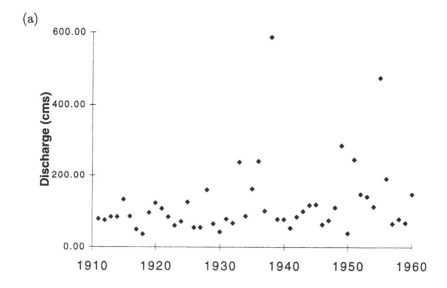

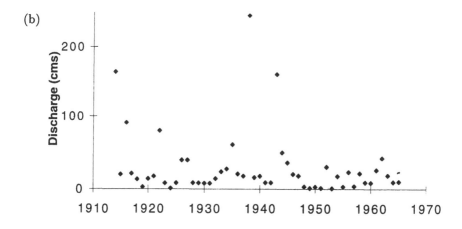

FIGURE 2 Maximum annual floods for (a) station 1-1805 on the Middle Branch of the Westfield River, Goss Heights, Massachusetts and (b) station 11-0980 in the Arroyo Seco near Pasadena, California.

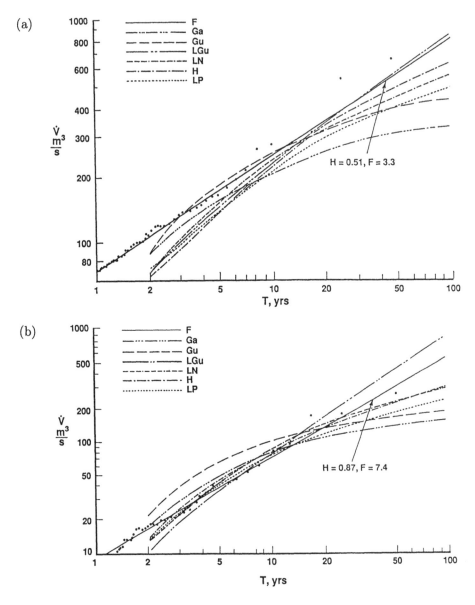

FIGURE 3 The points are the observed floods at the measuring stations during the periods considered (a) for Station 1-1805 and (b) for Station 11-0980. The peak discharge $\dot{V}$ is given as a function of recurrence intervals T. The scale-invariant (fractal) prediction, F, is compared with the six statistical predictions given by Benson[ ] two-parameter gamma (Ga), Gumbel (Gu), log-Gumbel (LGu), lognormal (LN), Hazen (H), and log-Pearson type III (LP). For the fractal correlation (F) the corresponding values of H and F are given.

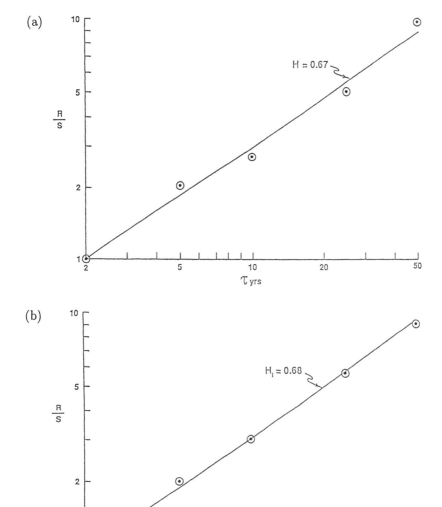

FIGURE 4 The rescaled range (R/S) for several intervals T. (a) Station 1-1805. (b) Station 11-0980. The correlations are with Eq. (9) and the Hurst exponents H_1 are given.

TABLE 1   Values of the Hausdorff measure H, fractal dimension D, flood intensity factor F, and Hurst exponent $H1_1$ for the ten benchmark stations

Station	River	(State)	H	D	F	H_1
1-1805	Westfield	(MA)	0.513	1.39	3.26	0.67
2-2185	Oconee	(GA)	0.540	1.46	3.47	0.72
5-3310	Mississippi	(MN)	0.470	1.53	2.95	0.72
6-3440	Little Missouri	(WY)	0.520	1.48	3.31	0.72
6-8005	Elkhorn	(NE)	0.540	1.46	3.47	0.67
7-2165	Mora	(NM)	0.630	1.37	4.27	0.73
8-1500	Llano	(TX)	0.719	1.28	5.24	0.70
10-3275	Humboldt	(NV)	0.616	1.38	4.13	0.66
11-0980	Arroyo Seco	(CA)	0.870	1.13	7.40	0.68
12-1570	Wenatchee	(WA)	0.310	1.69	2.04	0.72

However, the results indicate that there is considerable variation of F but very little variation in H_1. A simple explanation is that the former is sensitive to the Noah effect while the latter is sensitive to the Joseph effect. The relative scaling of floods is sensitive to the skewness of the statistical distribution but is not sensitive to the persistence of flows or floods. An important conclusion is that R/S analysis is not relevant to flood-frequency hazard assessments.

Many statistical distributions have been applied to historical records of floods. Benson[1] has given six statistical correlations for each of his ten benchmark stations. His results for the two-parameter gamma (Ga), Gumbel (Gu), log-Gumbel (LGu), lognormal (LN), Hazan (H), and log-Pearson type III (LP) are given in Figure 3(a) for station 1-1805 and in Figure 3(b) for station 11-0980. For large floods the fractal prediction (F) correlates best with the log-Gumbel (LGu) while the other statistical techniques predict longer recurrence time for very serious floods. The fractal and log-Gumbel are essentially power-law correlations whereas the others are essentially exponential.

While the ten benchmark stations provide a basis for comparing statistical approaches, they hardly made a convincing case that fractal statistics are preferable to alternatives. A principal difficulty is the relatively short time span over which reliable records have been collected. In order to try to overcome this difficulty, we have analyzed a large number of records and superimposed the results.

Our systematic study has considered 41-year records for 1009 USGS streamflow stations that are unaffected by flood control projects. A digitized record of daily mean discharge for the 40 years has been given for each station. The locations of the stations are given in Figure 5.

The best-fit straight lines have been obtained for the 1009 station records considered. The results will be divided into the 18 hydrologic regions of the conterminous United States illustrated in Figure 6. We first consider the quality of the fit of the data to the fractal relation (2). The ratios of the measured peak flows to the value predicted by the best fractal fit are determined for periods of $T = 1$ yr (fourtieth largest flood in the series), 5 yrs (eighth largest flood), 10 yrs (fourth largest flood), 20 yrs (second largest flood), and 40 yrs (largest flood). Results for the 111 stations in hydrologic region 3 are given in Figure 7(a), for the 123 stations in region 7 in Figure 7(b), for the 18 stations in hydrologic region 14 in Figure 7(c), and for the 100 stations in hydrologic region 17 in Figure 7(d). These results are typical of the 18 hydrologic regions. If all fits were perfect, then all data points would be unity. As expected, the scatter is greatest for the largest floods ($T = 40$ yrs) since they are based on a single data point. In general the scatter is quite small, a standard deviation of less than 5%.

FIGURE 5   Distribution of the 1009 stations that have been analyzed.

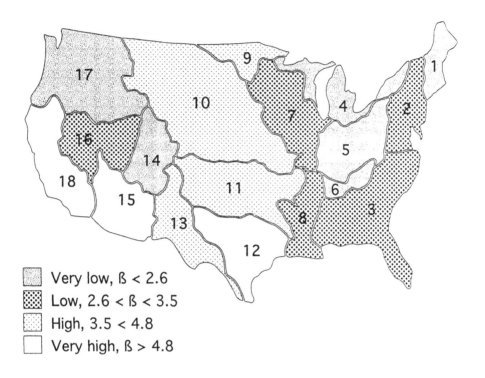

Very low, ß < 2.6
Low, 2.6 < ß < 3.5
High, 3.5 < 4.8
Very high, ß > 4.8

FIGURE 6   Hydrologic regions of the conterminous United States.[14] The shading gives the values of the flood frequency factor for the regions.

It is also of interest to compare the fractal fits for the stations within a hydrologic region. For each station the flow associated with the 10-year flood $\dot{V}_m(10)$ was normalized by the drainage area upstream of the station. The straight-line correlations for the 111 stations in hydrologic region 3 are given in Figure 8(a), for the 123 stations in hydrologic region 7 in Figure 8(b), for the 18 stations in hydrologic region 14 in Figure 8(c), and for the 100 stations in hydrologic region 17 in Figure 8(d). If all stations in a hydrologic region had the same flood intensity factor F and if peak flows were simply proportional to upstream drainage area, then all plots for the hydrologic region would lie on a common line. There is some variability in the flood intensity factor F (slope) and from one to two orders of magnitude variability in the normalized flows.

The mean value of the flood intensity factor for each of the eighteen hydrologic regions is given in Table 2. The standard deviation for each region is also given. Regional variations of the flood intensity factor F are clearly illustrated in this table. They are also illustrated in Figure 6 with very high ($F > 4.8$), high ($3.5 < F < 4.8$), low ($2.6 < F < 3.5$), and very low ($F < 2.6$) regions illustrated. Very high regions are the arid southwest and the western Gulf Coast (Texas). The latter is relatively arid and prone to hurricanes. The lowest value is the Pacific northwest with a strong maritime climate. Also very low are New England and the Great Lakes regions. The flood intensity factor appears to vary systematically with climate.

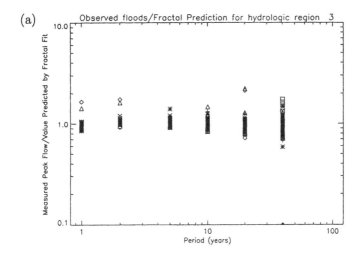

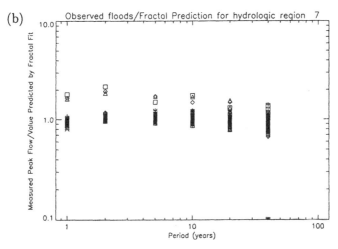

FIGURE 7   Ratio of the observed peak daily discharge to the value predicted by the fractal fit to the data as a function of the assigned period (a) for the 111 stations in region 3, (b) for the 123 stations in region 7, (continued)

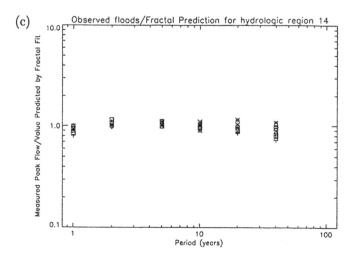

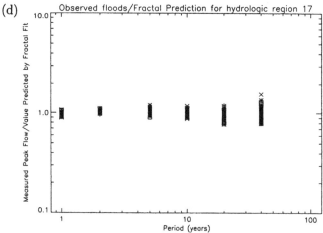

FIGURE 7   (continued) (c) for the 18 stations in region 14, and (d) for the 100 stations in region 17.

In some cases the standard deviations of the values of F in a hydrologic region are large. The largest standard deviation is for region 18. In Figure 9, the values of F for the 39 stations in this region (principally California) are plotted as a function of upstream drainage area A. For this purpose the region has been divided into three subregions: southern and central coastal California (SC), the Sierra Nevada (SN), and Northern California (NC). It is seen that the SC values are systematically high whereas the SN and NC values are systematically low. In general it is found that the flood intensity factor F is low if a substantial fraction of the drainage area has a winter snow pack. This is consistent with low values for the Sierra Nevada region. Northern California has a maritime climate similar to western Oregon and Washington and the law values of F are expected.

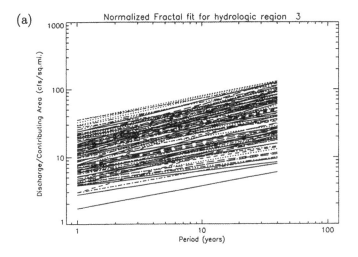

(a)

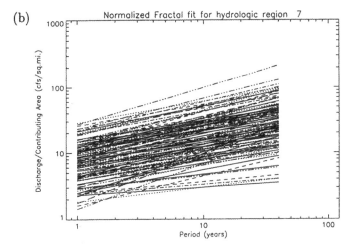

(b)

FIGURE 8   Fractal fits of the normalized flood frequency data (a) for the 111 stations in region 3, (b) for the 123 stations in region 7, (continued)

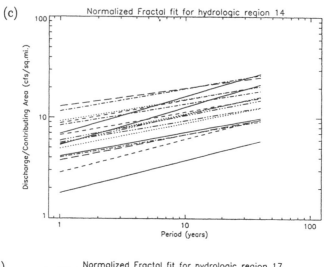

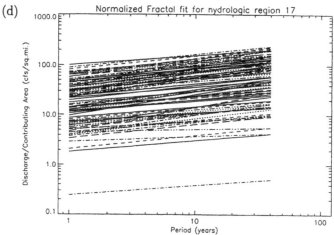

FIGURE 8 (continued) (c) for the 18 stations in region 14, and (d) for the 100 stations in region 17.

CONCLUSION

Historical flood-frequency records have been examined to determine whether fractal (power law) statistics are applicable. Although it must be recognized that the relatively short duration of historical records restricts the validity of conclusions, nevertheless, quite good agreement is obtained between fractal statistics and observations for 10 benchmark stations and for 1200 other stations in the United States.

The basic question in terms of flood hazard assessment is whether extreme floods decay exponentially in time or as a power law. If the power-law behavior is applicable, then the likelihood of severe floods is much higher and more conservative designs for dams and land use restrictions are indicated.

For fractal behavior the ratio of the 10-year to the 1-year flood F is also the ratio for the 100-year to the 10-year flood and the ratio of the 1000-year flood to the 100-year flood. We find large regional variations in values of F. In arid regions such as the southwestern United States, the values of F are nearly three times the values in more temperate regions such as the northwestern and northeastern corners of the country. Small values of F are also found if upstream drainage areas have large snow packs.

The relevance of R/S analysis to flood-frequency forecasting has also been addressed. For the ten benchmark stations we find the Hurst exponent to be $H_1 = 0.7 \pm 0.03$. This value indicates moderate persistence for the floods but also shows

TABLE 2 Average values and standard deviations of the flood intensity factor F for the 18 hydrologic districts

Hydrologic Regions	F	Standard Deviations	Number of Stations
1	2.369	0.377	54
2	2.998	1.313	147
3	2.758	0.617	111
4	2.183	0.289	57
5	2.396	0.509	129
6	2.505	0.324	38
7	2.782	0.738	123
8	3.021	0.979	22
9	4.7	1.586	13
10	3.557	1.677	64
11	3.897	1.801	46
12	4.848	1.559	13
13	4.104	2.121	14
14	2.283	0.51	18
15	6.066	1.08	11
16	2.778	0.752	10
17	2.076	0.357	100
18	5.134	2.4	39

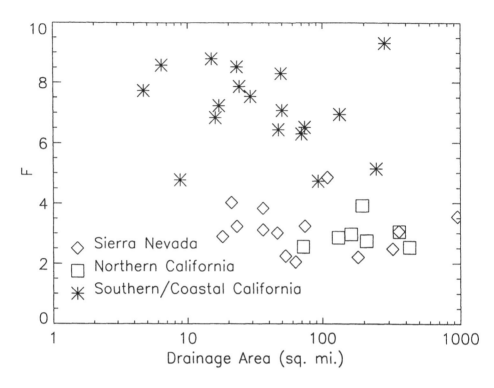

FIGURE 9   Flood intensity factors as a function of upstream drainage areas for the 39 streamflow stations in hydrologic region 18 (California). Data are given for three subregions: southern and central coastal California (SC), the Sierra Nevada (SN), and Northern California (NC).

that determinations of Hurst exponents are not useful for flood hazard assessments. The Hurst exponent H_1 does not correlate with the fractal flood parameter F. In the terms introduced by Mandelbrot and Wallis,[7] the Hurst exponent H_1 is sensitive to the Joseph effect or persistence of events whereas the fractal flood parameter F is sensitive to the Noah effect or skewness of the statistical distributions of floods.

A primary objective of this paper has been to show that the two-parameter fractal distribution is in as good agreement with observations, on average, as empirical laws with three or more free parameters. However, a more fundamental objective is to argue that there is a physical basis for preferring the scale-invariant fractal relation to empirical alternatives that involve one or more scaling parameters. A valid question is whether all the complexities associated with floods could be consistent with one statistical law. Physical processes associated with floods include: (1) The amount of rainfall produced by the storm or storms in question, (2) the upstream drainage area, (3) the saturation of the soil in the drainage area, (4) the topography, soil type, and vegetation in the drainage area, and (5) whether snow melt is involved. In addition, dams, stream channelization, and other man-made modifications can affect the severity of floods. However, a variety of complex natural phenomena have been shown to obey fractal statistics[11] including earthquakes and volcanic eruptions.

An important aspect of fractal statistics is the relation of fractional noises to fractional walks. In general, fractional noises do not have power law distributions of extreme values although the walks do. If river flows are equivalent to noises (reservoir volumes equivalent to walks), why should the extremes of river flows (floods) have a power law distribution? But the river flows themselves represent the addition of individual rain events (storms); thus it is not unreasonable to consider river flows as walks rather than noises. As walks, a power law distribution of extreme values (floods) would be expected.

REFERENCES

1. Benson, M. A. "Uniform Flood-Frequency Estimating Methods for Federal Agencies." *Water Resour. Res.* **4** (1968): 891–908.
2. Feder, J. *Fractals.* New York: Plenum Press, 1988.
3. Green, A. R. "Magnitude and Frequency of Floods. Part 1-A." Water Supply Paper 1671, U.S. Geol. Survey, 1964, pp. 212–213.
4. Hurst, H. E. "Long-Term Storage Capacity of Reservoirs." *Am. Soc. Civil. Eng. Trans.* **116** (1951): 770–779.
5. Hurst, H. E. "Methods of Using Long-Term Storage in Reservoirs." *Inst. Civil Eng. Proc.* Part 1, **5** (1956): 519–590.
6. Hurst, H. E., R. P. Black, and Y. M. Simaika. *Long-Term Storage.* London: Constable, 1965.
7. Mandelbrot, B. B., and J. R. Wallis. "Noah, Joseph, and Operational Hydrology." *Water Resour. Res.* **4** (1968): 909–918.
8. Mandelbrot, B. B., and J. R. Wallis. "Computer Experiments with Fractional Gaussian Noises." Parts I, II, III. *Water Resour. Res.* **5** (1969): 228–267.
9. Mandelbrot, B. B., and J. R. Wallis. "Some Long-Run Properties of Geophysical Records." *Water Resour. Res.* **5** (1969): 321–340.
10. Mandelbrot, B. B., and J. R. Wallis. "Robustness of the Rescaled Range R/S in the Measurement of Noncyclic Long Run Statistical Dependence." *Water Resour. Res.* **5** (1969): 967–988.
11. Turcotte, D. L. *Fractals and Chaos in Geology and Geophysics.* Cambridge: Cambridge University Press, 1992.
12. Turcotte, D. L., and L. Greene. "A Scale-Invariant Approach to Flood-Frequency Analysis." *Stochastic Hydrol. Hydraul.* **7** (1993): 33–40.

13. U.S. Water Resources Council. "Floods Flow Frequency." Bulletin 17B, Hydrology Committee, Washington, DC, 1981.

14. Wallis, J. R., D. P. Lettenmaier, and E. F. Wood. "A Daily Hydroclimatological Data Set for the Continental United States." *Water Resour. Res.* **27** (1991): 1657–1663.

15. Young, I. E., and R. W. Cruff. "Magnitude and Frequency of Floods." Part 11. Water Supply Paper 1685, U.S. Geol. Survey, 1967, pp. 714–715.

Vijay K. Gupta† and Ed Waymire‡
†Center for the Study of Earth from Space/CIRES, Department of Geological Sciences, University of Colorado, Boulder, CO 80309
‡Departments of Mathematics and Statistics, Oregon State University, Corvallis, OR 97331

Multiplicative Cascades and Spatial Variability in Rainfall, River Networks, and Floods

Our objective here is to provide an overview of recent research on scaling invariance in the spatial variability of rainfall, river networks, and floods. We briefly review the developments within the last decade in the mathematical theory of cascades, and the applications of the cascade theory to the river network structure and to spatial rainfall. These applications in conjunction with empirical analyses of spatial variability of floods suggest possible physical hypotheses regarding scaling invariance in the regional structure of floods. To this end we set out to identify and compute basic "scaling structure functions" associated with each of the following: (i) river networks, (ii) rainfall, and (iii) floods. The idea to quantify properties of these systems in terms of their scaling structure functions is not unlike their use in the statistical theory of turbulence to describe the hierarchy of structure observed there. The difficulty is, of course, in identifying the scaling regularities. Nonetheless, as will be seen, the success has been quite remarkable. The scaling structure functions ultimately provide a basis on which various connections can be examined. As a result one gets the picture of a newly developing theoretical framework for synthesizing spatial variability in rainfall and in river networks being used to understand, interpret, and test various physical hypotheses regarding the spatial structure of river runoff and floods.

1. INTRODUCTION

A generic problem in the hydrologic sciences is to make extrapolations to space and time domains in which no data exist and no direct measurements of processes can be made. Consequently predictions require an understanding of aggregation and disaggregation of processes across a broad range of spatial and temporal scales. On the physical side, in this regard, the processes of precipitation, runoff, soil moisture, evapotranspiration, etc. have attracted the most attention in recent years. These particular processes are judged to be most relevant to applications in climate models, measurements of rainfall from radar and satellite, and prediction of floods from ungauged river basins to mention a few common ones. For the most part these are multiple-scale phenomena and one seeks to identify scale ranges over which the hypothesis of scaling invariance can be formulated and tested. Scaling invariance over a prescribed range of scales means that the dominant features of a system cannot be inferred from a snapshot of it since they remain invariant as the size or the scale of the system is changed.

Scaling invariance is described by the so-called "scaling exponents" given that the appropriate scaling structure functions can be identified. The most widespread use of this approach in quantifying large systems possessing hierarchial structure has occurred in statistical physics and statistical theories of turbulence. The corresponding approach is more recent in the context of the hydrologic sciences. Nonetheless, two dominant structural patterns observed in the natural processes of interest to us here are branching and clustering, and the nature and consequences of branching in river networks and of clustering in spatial rainfall at different scales of resolution have been a major research theme in the earth science literature over the past half century. Among a variety of mathematical constructs, e.g., percolation and cellular automaton models, renormalization groups, Levy stable, Brownian and fractional Brownian noises, multiplicative cascades, etc., that have been developed to compute the structure functions in different contexts, the theory of cascade measures is most appropriate to the present context. Our objective in this chapter is to review some of the recent developments which suggest ways in which the theory of cascades has the potential to unify disparate processes in river basins and thereby fundamentally aid our approach to river basin hydrology in the years ahead.

This chapter is organized as follows. Section 2 gives a brief history and an informal review of the mathematical theory of cascades. Two simple examples consisting of the deterministic binomial cascade and the random beta model are used to illustrate the theory. In the first reading, the reader may wish to skip this section and go directly to Section 3. It is divided into three main subsections covering channel networks, rainfall, and floods. A brief historical perspective of the literature is given in each of these three subsections. We first review the newly developing theory of self-similar river network topology and the theory of deterministic cascades, from the point of view of partitioning a basin into successively smaller subbasins. This is followed by a review of how the theory of random cascades is being used to describe spatial rainfall variability and clustering across a broad range of scales. Finally, we review recent developments in the spatial analyses of annual peak flow data which show that the structure function of rainfall generated floods is qualitatively similar to that of spatial rainfall in "large river basins." By contrast, the structure functions of snow-melt-generated floods, and floods in "small basins" seem to be similar to each other, and are different from that of floods in large basins. After reading Section 3 the reader can go to Section 2 to develop some background in the theory of cascades. Finally, Section 4 discusses how the theories of rainfall and river networks would combine for explaining the empirically observed scaling structure of floods, and in developing a suitable theory of river runoff and floods.

---

2. CASCADE PRELIMINARIES

Our focus in this section is on a class of mass or energy distributions on a region $X \subseteq R^d$ represented by measures possessing a natural multiplicative structure, referred to as a *cascade structure*. For examples, X may denote d-dimensional Euclidean space, the d-dimensional cube $[0,1]^d$, or some other metric space of interest. This class of measures owes its origins to early ideas of Kolmogorov[31,32] on statistical turbulence theories and continues to play a prominent role in both turbulence theory and data analysis; e.g., and see She and Leveque,[33] Dubrulle,[7] She and Waymire[34] for a very recent correction to Kolmogorov's 1962 lognormal hypothesis by a log-Poisson distribution. In the case of deterministic cascades one also finds applications to the structure of certain invariant measures of chaotic dynamical systems; see Feigenbaum et al.[12] However, as will be explained in this chapter, the scope of applications of both random and deterministic cascade measures extends naturally to hydrologic phenomena possessing a hierarchy of scales; also see Feder[11] in this regard.

Recall that by a measure μ defined on a region X one simply means an assignment of nonnegative numerical values (mass, energies, etc) $\mu(A)$ to certain subregions A of X, called the *measurable subsets*,[1] such that $\mu(\emptyset) = 0$ and for any disjoint sequence $A_1, A_2, \ldots$ one has the *additivity condition*:

$$\mu(A_1 \cup A_2 \cup \cdots) = \mu(A_1) + \mu(A_2) + \cdots. \tag{2.1}$$

---

[1]The collection $\mathcal{F}$ of those subregions A of X which are assigned mass $\mu(A)$ is required to be closed under sequential application of the set operations *union, intersection, complement;* this is the *sigmafield property* of $\mathcal{F}$.

The most important reference measure is that of d-dimensional volumes, denoted $\lambda_d(A)$, of subregions A of $\mathbf{R}^d$. This case is referred to as *d-dimensional Lebesgue measure*.

Cascade measures on the d-cube are a special class of mass distributions on $X = [0,1]^d$ defined recursively in terms of a natural hierarchial sequence of scales of resolution of the cube. For simplicity of the exposition let us first take $d = 1$ and consider the hierarchy of scales associated with the succesive binary partitions of the unit interval $X = [0,1]$ given by (see Figure 1) $n = 1 : \Delta(0) = [0,1/2), \Delta(1) = [1/2,1]$; $n = 2 : \Delta(00) = [0,1/4), \Delta(01) = [1/4,2/4), \Delta(10) = [2/4,3/4), \Delta(11) = [3/4,3/4,1]$; etc. So the binary resolution of X at the nth scale is given by the 2^n subintervals naturally coded-up as

$$\Delta(t_1 \ldots t_n) = \left[\sum_{i=1}^{n} t_i 2^{-i}, \sum_{i=1}^{n} t_i 2^{-i} + 2^{-n} \right) \tag{2.2}$$

for $t_i \in \{0,1\}$. Ternary, quartary, and more generally b-ary partitions, for a parameter $b = 2,3,4,\ldots$, are obtained similarly and can be coded-up in terms of b-ary expansions $\sum_{i \geq 1} t_i b^{-i}, t_i \in \{0,1,\ldots,b-1\}$ of the partition points. Similarly one may consider a b-ary partition of the d-cube and code up the cells in the case $d \geq 2$. For simplicity, return to $d = 1$ and the binary partition of $X = [0,1]$. The defining parameters of the model are furnished by nonnegative weights $W_0, W_1, W_{00}, W_{01}, W_{10}, \ldots, W_{t_1 t_2 \ldots t_n}, \ldots$, referred to as *cascade generators*. Given the generators, one distributes mass at resolution $n = 1$ according to the measure defined by

$$\mu_1(A) = \int_A Q_1(x) dx \tag{2.3}$$

where the density $Q_1(x)$ is piecewise constant with value W_0 on $\Delta(0) = [0,1/2)$ and W_1 on $\Delta(1) = [1/2,1]$; see Figure 1. For $n = 2$ this density is redistributed multiplicatively as follows. The value $Q_1(x) = W_0$ for $x \in \Delta(0) = \Delta(00) \cup \Delta(01)$ is redistributed as $Q_2(x) = W_{00}W_0$, for $x \in \Delta(00)$ and as $Q_2(x) = W_{01}W_0$, for $x \in \Delta(01)$. Similarly the value $Q_1(x) = W_1$ for $x \in \Delta(1) = \Delta(10) \cup \Delta(11)$ is redistributed as $Q_2(x) = W_{10}W_1$, for $x \in \Delta(10)$, and as $Q_2(x) = W_{11}W_1$, for $x \in \Delta(11)$. Iterating this process one obtains at the nth resolution a piecewise constant density $Q_n(x)$ having 2^n pieces of the multiplicative form

$$Q_n(x) = W_{t_1 \ldots t_n} W_{t_1 \ldots t_{n-1}} \cdots W_{t_1 t_2} W_{t_1}, x \in \Delta(t_1 t_2 \ldots t_n). \tag{2.4}$$

The corresponding cascade mass distribution is then defined at the nth resolution with $Q_n(x)$ serving as the density by

$$\mu_n(A) = \int_A Q_n(x) dx. \tag{2.5}$$

The cascade measure may now be defined by the *fine-scale-limit measure* $\mu_\infty(A) = \lim_{n\to\infty}\mu_n(A)$. While such limits may exist for a suitably large class of sets A, in general it is *NOT* the case the the limit measure will have a density; i.e., the limit measures are typically thinly supported with high degrees of intermittency. In particular for each x the density $Q_n(x)$ will typically die off to zero or explode to infinity as $n \to \infty$ so it is generally important to consider the integrated mass when taking limits.

In order to obtain fine scale limits of the type described above, one must impose certain *mass conservation* conditions on the cascade generators. Some naturally occuring conservation laws appear in the following subsections. However, beyond this the structure is quite general, allowing for both random and nonrandom variability in the cascade generators. The reader who is being introduced to cascades for the very first time may wish to consult one of the many expository references now available, for example, Evertsz and Mandelbrot[9] or Falconer.[10]

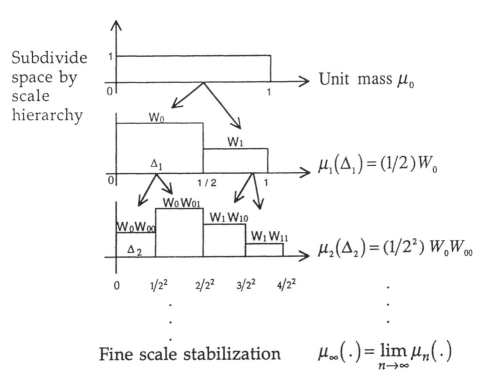

FIGURE 1   A schematic of mass distribution in a multiplicative cascade.

2.1 DETERMINISTIC MULTINOMIAL CASCADES

The deterministic binomial cascade on $X = [0,1]$ is obtained by a $(b = 2)$ binary partition with generators of the cascade density given by $W_0 = 2m_0, W_1 = 2m_1$ and subject to the mass conservation condition

$$m_0 + m_1 = 1. \tag{2.6}$$

In the definition of the mass density, note that the length scale is $1/2$ at the $n = 1$ level of resolution. Mass is then redistributed multiplicatively in the same proportions so that the generators for the iteration are simply of the form:

$$W_{t_1 t_2 \cdots t_n} = 2m_{t_n}, t_i \in \{0,1\}. \tag{2.7}$$

That is, the piecewise constant density at the nth resolution is given by

$$Q_n(x) = W_{t_1 \ldots t_n} W_{t_1 \ldots t_{n-1}} \cdots W_{t_1 t_2} W_{t_1} = 2^n m_{t_1} m_{t_2} \cdots m_{t_n}, x \in \Delta(t_1 t_2 \ldots t_n) \tag{2.8}$$

and the integrated mass of the nth level cell $\Delta(t_1 t_2 \ldots t_n)$ is

$$\mu_n(\Delta(t_1 t_2 \ldots t_n)) = m_{t_1} m_{t_2} \cdots m_{t_n}, \tag{2.9}$$

where each $m_{t_i} = m_0$ or m_1 depending on $t_i \in \{0,1\}$. In the case of this example one has the following rather special consistency equation

$$\mu_\infty(\Delta(t_1 \ldots t_n)) = \mu_n(\Delta(t_1 \ldots t_n)). \tag{2.10}$$

The special nature of this equation will become more clear from the considerations in section 2.2; see Eq. (2.14). As noted above, observe that, unless $m_0 = m_1 = 1/2$, either $Q_n(x) \to 0$ or $Q_n(x) \to \infty$ as $n \to \infty$ for each fixed x since there are only two parameter values m_0, m_1. However if $m_0 = m_1 = 1/2$, then $Q_n(x) \equiv 1$ and the resulting measure is simply Lebesgue measure.

The more general multinomial cascade is obtained exactly in the same way using a b-ary partition and cascade parameters $m_0, \ldots, m_{b-1}$ for the generators subject to the conservation condtion $m_0 + \cdots + m_{b-1} = 1$.

While the deterministic cascade arises naturally in connection with river network geomorphology, as described in section 3.1 and, as noted earlier, in the context of dynamical systems, other applications, e.g., rainfall distributions, require extension to *random generators*. A randomized version will be described next.

2.2 RANDOM MULTINOMIAL CASCADE

In this example let us again consider the binary partition with two parameters m_0, m_1 as in Eq. (2.1) but now, at each stage we independently randomly select from these two values for a generator value with equal probabilities. That is

$$W_{t_1 t_2 \cdots t_n} = \begin{cases} 2m_0, & \text{with probability } 1/2; \\ 2m_1, & \text{with probability } 1/2. \end{cases} \tag{2.11}$$

All else is the same as in Eq. (2.1), though now the values of the cascade measure are randomly varying. Note also that the conservation condition is now a *conservation of averages*. That is, the expected values

$$EW_{t_1 t_2 \cdots t_n} = 2m_0 \times \frac{1}{2} + 2m_1 \times \frac{1}{2} = 1 \tag{2.12}$$

are conserved, but, for example in 25% of the samples, one will have $W_0 = W_1 = 2m_0 \neq 1$, as a result of randomizing the selections.

2.3 BETA MODEL

Consider b-ary partitions of $X = [0,1]$. A simple but important choice for the generator distribution of a random cascade is that of independent and identically distributed generators distributed as

$$W = \begin{cases} b^\beta & \text{with probability } b^{-\beta}; \\ 0 & \text{with probabiility } 1 - b^{-\beta}. \end{cases} \tag{2.13}$$

One might guess that if the parameter β is very large, then the probability of a zero factor in the multiplicative cascade will be close to one and hence the limiting measure will certainly assign mass zero to all subregions. On the other hand, for small positive values of β the occurrence of a zero generator in the redistribution of mass is unlikely and one may expect to get a nondegenerate mass distribution in the limit. The critical value of β for nondegeneracy is given by the equation $\beta_c = 1$. Note that b values of the cascade generators must be randomly generated in passing from level n to level $n+1$. The corresponding expected number of nonzero generator values is then $bb^{-\beta} = b^{1-\beta}$. The *critical value* of β is such that the limit cascade is degenerate if this expected number is one or smaller, i.e., $\beta \geq 1 = \beta_c$, while there is a positive probability that the limit cascade is nondegenerate if $\beta < 1$. In the case of this example one has

$$\mu_\infty(\Delta(t_1 \ldots t_n)) = Z_\infty(t_1 \ldots t_n)\mu_n(\Delta(t_1 \ldots t_n)) \tag{2.14}$$

where the prefactor $Z_\infty(t_1 \ldots t_n)$ is statistically independent of the nth-level mass $\mu_n(\Delta(t_1 \ldots t_n))$ and is distributed as the total mass $\mu_\infty(X)$. Observe that the case of the deterministic cascade in section 2.1 is made special by the fact that the corresponding prefactor is identically one; see Eq. (2.10). Other critical phenomena associated with the beta model arise in connection with the intermittency structure of the limit mass which can be calculated in terms of connectivity properties of limiting mass given that it forms; e.g., see Chayes, Chayes, and Durrett[4] and Mandelbrot.[35]

More generally, one may generate nonnegative values according to any independent and identically distributed choice of the generators, e.g., lognormal, Gamma, log-Poisson, etc., so long as a suitable average conservation condition is satisfied (or else the limits will not exist); see Kahane[27] and Waymire and Williams.[36,37]

2.4 FINE-SCALE STRUCTURE AND SCALING EXPONENTS

In this section we will briefly consider the fine-scale structure of random cascades. We summarize here two standard approaches to the computation of fine-scale structure in terms of certain multifractal scaling exponents and structure functions.

As noted earlier the typical behavior is for the cascade densities $Q_n(x)$ to either die off or diverge to infinity in the fine-scale limit. So while there is no value to trying to identify a limiting density, one may investigate the rate at which density dies off or explodes as a function of scale. For example, consider the deterministic binomial cascade discussed above. One has for the leftmost interval at the length scale resolution $l_n = 2^{-n}$

$$\mu_\infty[0, l_n) = \mu_\infty(\Delta(00\cdots 0)) = m_0^n = l_n^{\alpha_0} \qquad (2.15)$$

and for the rightmost subinterval

$$\mu_\infty[1 - l_n, 1] = \mu_\infty(\Delta(11\cdots 1)) = m_1^n = l_n^{\alpha_1} \qquad (2.16)$$

where

$$\alpha_i = -\log_2 m_i, \quad i = 0, 1. \qquad (2.17)$$

Similarly, if $\Delta(t_1 \ldots t_n), n = 1, 2, \ldots$ is a nested sequence of intervals shrinking to a point x, then, in the fine-scale limit $l_n = 2^{-n} \to 0$ as $n \to \infty$, one has

$$\mu_\infty(\Delta(t_1 \ldots t_n)) \sim l_n^{\alpha(x)} \qquad (2.18)$$

in the sense that

$$\alpha(x) = \lim_{l_n \to 0} \frac{\log \mu_\infty(\Delta(t_1 \ldots t_n))}{\log l_n}. \qquad (2.19)$$

The parameter $\alpha(x)$ is a *local Holder exponent* or *order of the singularity* at the location x.

Since the Holder exponents measure the local strengths of the singularities in the limit cascade, it is of interest to determine "how much" of the space is occupied by singularities of a given order α. A convenient measure of the "size" of the region of growth of order α is in terms of another scaling exponent $f(\alpha)$ giving the Hausdorff dimension of that region. This exponent function is often referred to as the *singularity spectrum*. In the case of the binomial cascade one has (e.g., see Falconer,[10] Evertsz and Mandelbrot[9]):

$$f(\alpha) = - \left\{ \frac{A - \alpha}{A - a} \log_2 \left(\frac{A - \alpha}{A - a} \right) + \frac{\alpha - a}{A - a} \log_2 \left(\frac{\alpha - a}{A - a} \right) \right\}, \qquad (2.20)$$

where A, a are the largest and smallest Holder exponents.

Another related computation of the fine-scale structure of cascades may be made in terms of the behavior of a *cascade partition function* defined in analogy with statistical mechanics, or the *cascade structure function* from statistical turbulence defined as follows. For each $q \in \mathbf{R}$ define

$$S_{l_n}(q) = \sum_{(t_1, \ldots, t_n)} \mu_\infty^q (\Delta(t_1 \ldots t_n)) \qquad (2.21)$$

at the scale of resolution $l_n = 2^{-n}$. In the case of the binomial cascade one has a *scaling* structure function

$$S_{l_n}(q) = (m_0^q + m_1^q)^n = l_n^{-\log_2(m_0^q + m_1^q)} = l_n^{\tau(q)} \qquad (2.22)$$

where, in general,

$$\tau(q) = - \lim_{l_n \to 0} \frac{\log S_{l_n}(q)}{\log l_n} \qquad (2.23)$$

is the *structure function exponent*. For the binomial cascade one has

$$\tau(q) = \log_2(m_0^q + m_1^q). \qquad (2.24)$$

In this case one may check that $f(\alpha)$ and $\tau(q)$ are related as *Legendre transform pairs*. As will be illustrated in section 3.2, for certain hydrologic applications the computation of these exponents are also required for the case of random cascades. Fortunately, the corresponding relationship holds more generally for a large class of random cascades as well; see Holley and Waymire.[24]

---

3. APPLICATIONS TO RIVER BASIN HYDROLOGY

The river basins represent a natural unit of landscape organization. Worldwide the size of basins as measured by their drainage area A, varies over eight orders of magnitude. As the size of a basin increases, various hydrologic processes in a river basin, be it river runoff and floods, or soil moisture, or evapotranspiration, are strongly influenced by two "boundary conditions." The first is the space-time variability in precipitation, which includes both rainfall and snowfall, and second is the spatial variability in landforms which also includes variability in vegetation. This review includes the relevance and application of cascade theory to the two boundary processes, i.e., space-time rainfall and landforms via channel networks. With respect to river basin hydrology, we have chosen to focus on the recent research on scaling invariance in the spatial variability of peak flows or floods (Gupta and Waymire[20]; Gupta, Mesa, and Dawdy[16]). This choice is motivated by the fact that floods have a direct connection with rainfall and snow melt. In addition, floods are also viewed as "channel forming discharges" and therefore have a direct connection with channel networks and landforms. These empirical relations are known as "hydraulic geometry" in the literature (Leopold, Wolman, and Miller[38]). We will explain how applications of cascade theory to rainfall suggest an entirely new way to interpret the presence of scaling invariance in the empirical flood data (Gupta and Dawdy[17]). The role of cascades will then be discussed in developing a space-time theory of river runoff and floods in Section 4.

3.1 CHANNNEL NETWORKS AND DETERMINISTIC CASCADES

INTRODUCTION AND A BIT OF HISTORY. The period from 1945 to 1965 witnessed the discovery of a variety of empirical relationships involving drainage network bifurcation and geometric structure. These relationships are based on the Horton-Strahler (H-S) ordering scheme. Well-known among them are the so-called *Horton laws of stream numbers, lengths and areas* (Smart[39]), and *Hack's law* (Hack, 1957). These relationships were then extended to hydraulic-geometric variables, which included annual peak flows, channel slopes, widths, depths etc.[38] Shreve showed that the Horton laws and many other empirical relationships can be derived from the hypothesis of "topologic randomness." The period from the late 1960s to almost late 1980s was dominated by Shreve's random model. Various tests were carried out to test the hypothesis of topologic randomness as the null hypothesis against observations. From the late 1970s to the late 1980s, it also included the development of a sizable hydrology literature on the geomorphologic unit hydrograph (GIUH), which was devoted to establishing the extent to which the network structure can be used in determining the flood response, or the hydrograph, of a basin; see the review by Gupta and Mesa.[15]

Starting in the late 1980s to early 1990s, attempts were made to extend the theory of channel networks to include hydraulic geometry. Investigations of the spatial structure of annual peak flows (Gupta and Waymire[20]) and channel slopes (Gupta and Waymire[19]) revealed quite clearly that the ideas of scale invariance and statistical self-similarity are of fundamental importance in these generalizations of the theory of channel networks and in interpreting empirical observations. Recently, Peckham[40,41] has argued that Horton's ordering scheme is of basic significance in formulating a precise answer to the following question: In what sense can two river networks of different sizes said to be topologically and geometrically similar to one another? In the link-based enumeration of a river network introduced by Shreve, it is not at all clear how one can think of two basins as being topologically similar! The Horton ordering led to developing a theoretical framework of self-similar trees (SST) by Tokunaga.[42] Even though this theory has not been widely known in the literature, its generalizations lead to a powerful predictive theory.[40] In addition, a somewhat surprising combinatorial connection with early statistical mechanics of Riddell and Uhlenbeck[43] was recently uncovered by Maxwell,[44] which renders a systematic computation of stream numbers of all orders for SSTs. SSTs not only exhibit many different empirical "laws" obeyed by river networks, but also make it possible to develop and test the notions of dynamical and statistical self-similarity in river networks within a Hortonian framework; see Peckham.[41] We will briefly describe the SST framework, followed by a computation of its structure function via connections to deterministic cascades.

SELF-SIMILAR TREES AND THEIR STRUCTURE FUNCTIONS. We begin our discussion with the following question. How does a branching river network partition a three-dimensional region enclosed by a drainage divide, into successively smaller regions known as subbasins? This is a very important problem and some of the early ideas on how a network fills space go back to Horton; see Jarvis and Woldenberg[25] for many classic papers on this topic. In the present context, we are interested in the question of how an SST partitions a drainage basin. Clearly, a cascade theory with statistically independent random generators cannot be applied to this problem since the total area of a basin is conserved in the partitioning. This strong conservation will correlate the cascade generators if there is any randomness. Therefore, it restricts the application of cascade theory to the class of either deterministic or statistically correlated cascade generators. For simplicity and illustration, we only consider deterministic cascades in this paper.

Recall from Section 2 that in a cascade construction, branching is governed by an SST. We consider a subclass called *recursive replacement trees*. Unlike river networks, these trees are not binary. Nonetheless, this choice illustrates an important connection between SSTs and multiplicative cascades, and makes precise the notion of the structure function of a river network. Some well-known special cases of recursive replacement trees are the idealized Peano network (Marani, Rigon, and Rinaldo[45]), and the tree considered by Mandelbrot-Viscek.[46]

The SSTs are constructed in a manner which is very similar to constructing many well-known fractal curves, e.g., Koch curve, Cantor set, etc. It is based on the so-called "Tree Generators," which are different from cascade generators, so the reader is reminded of the difference between these two types of generators. A tree generator, denoted by $T_{\omega,k}$, represents the number of side tributaries of order $1 \le k < \omega$ entering a tributary of order ω. *Topologic self-similarity* is defined by the condition that $T_{\omega,k}$ is the same for different ordered streams, i.e., $T_{\omega,\omega-k} = T_k$. The generators of recursive replacement trees are specified by the branching number b, and the number n of nodes in a stream as

$$T_1 = (b-1)(n-1), \quad T_k = (b-1)n^2(n+1)^{k-2}, \ k \geq 2. \tag{3.1}$$

The recursive structure of an SST gives rise to a recursion equation for the variable N_ω, which denotes the streams of order ω in a network. The recursion equation is used to test for the validity of the well-known Horton law of stream numbers, which is obtained as (Peckham[40,41])

$$\lim_{\omega \to \infty} \frac{N_\omega}{N_{\omega+1}} = R_b = nb + 1 \tag{3.2}$$

where R_b is known as the bifurcation ratio. Two other important quantities predicted by an SST are the "link number," R_c, denoting the asymptotic limit of the ratio of number of links in a stream of order $\omega + 1$ to that in a stream of order ω, and the "topologic fractal dimension," D. They are given by,

$$R_c = n + 1, \quad D = \frac{\log R_b}{\log R_c}. \tag{3.3}$$

For the Peano tree, plugging in the values of b and n into above formulae gives $R_c = 2, R_b = 4, D = 2$. Interestingly, these values are the same as predicted by the random topology model, yet the recursive replacement trees differ fundamentally from the the topology of an "Average Shreve tree," insofar as the latter is a binary SST and its generators are given by $T_k = 2^{k-1}, k \geq 1$.

In order to compute the structure function of a SST, we will define the notion of the *width function* of a network. Within the last two decades, various attempts to seek the connections between river runoff and channel network geometry and topology have identified two important functions. The first is known as the "width function" (Kirby[30]), and the second "the link concentration function" (Gupta, Waymire, and Rodriguez-Iturbe[22]). The term "link" refers to the segment of a channel between two consecutive junctions. The width function $\{N(x) : 0 \leq x \leq L\}$ denotes the number of links at a flow distance from the basin outlet divided by the total number of links, where L is the length of the longest channel in a network. If the topology of a network is specified by a probability model, e.g., the random model, then the width function can be viewed as a stochastic process indexed by the spatial distance x. This formulation was used by Troutman and Karlinger[47] and Waymire[48] to derive results concerning the width function for the random topology model. However, within the context of a deterministic SST we do not need to concern ourselves with this added complication of random topologies.

To illustrate the computation of the structure function of a recursive replacement tree, we consider the Peano tree. Take a channel of unit length at the initial stage of construction. The width function is the unit mass distributed uniformly over $[0,1]$. At the first stage this channel is subdivided into half, and two side tributaries are added at the node in the middle, as shown in Figure 2(a). The width function can be written as

$$N(x) = \begin{cases} 1/4 & \text{if } 1 \le x \le 1/2, \\ 3/4 & \text{if } 1/2 < x \le 1. \end{cases} \tag{3.4}$$

At the second stage of construction, each link of length $1/2$ is subdivided into half, and two side tributaries are added at each node in the middle, as the width function shown in Figure 3(b) can be expressed as

$$N(x) = \begin{cases} 1/16 & \text{if } 1 \le x \le 1/4, \\ 3/16 & \text{if } 1/4 < x \le 1/2, \\ 3/16 & \text{if } 1/2 < x \le 3/4, \\ 9/16 & \text{if } 3/4 < x \le 1. \end{cases} \tag{3.5}$$

The two width functions corresponding to the two stages of construction are plotted in Figure 2(a),(b). It is clear from this figure that the width function can be viewed as a binomial cascade described in section 2.1 with an initial uniform mass density and the cascade parameters $m_0 = 1/4, m_1 = 3/4$. The structure function exponent is given by the expression in Eq. (2.24). This construction extends to the broad class of recursive replacement SSTs and is described by Gupta, Peckham, Veitzer, and Waymire.[17]

Before closing this section, we wish to remark that our ongoing research also shows that the asymptotic mass distribution for SST topologies outside the class of recursive replacement trees can be quite different than the singular behavior described above. For example, the asymptotic mass distribution for the average Shreve tree has recently been shown to converge to a uniform distribution.[49]

(a)   (b)

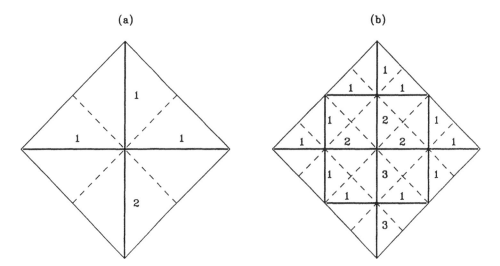

FIGURE 2   A schematic of two stages of Peano tree construction (from Marani et. al.[45]

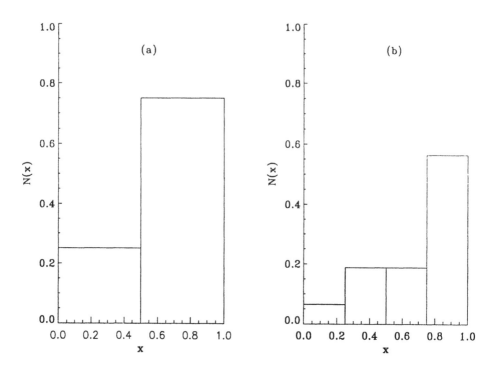

FIGURE 3   Width functions corresponding to the two stages of Peano tree construction.

3.2 SPATIAL RAINFALL AND RANDOM CASCADES

INTRODUCTION AND A BIT OF HISTORY. During the 1950s and 1960s a central theme
of rainfall research in hydrology was to fit parameters of various time series models
to point rainfall measurements on time scales ranging from hourly, daily, monthly,
and yearly. Examples of time series models include mth-order wet/dry Markov chain
models, renewal sequences, and moving average models; see Katz,[26] and Waymire
and Gupta.[18] As attention turned to more physically based approaches to take into
account observed clustering in space and time, this theme was further explored into
the following decades using compound Poisson and compound Neyman-Scott time
series models; see Rodriguez-Iturbe.[50] The lesson learned was that up to second
order, moment characteristics could be "reasonably well reproduced by a variety
of models at a given scale." Better understanding of the temporal evolution would
require the physical structure of storm events which is furnished by spatial observations.

   The hierarchical structure of spatial rainfall fields takes the form of clusters of
high-intensity rain cells embedded in clusters of lower intensity regions, called small

mesoscale areas (SMSA), which are in turn embedded in rainbands of identifiably lower intensity, called large mesoscale areas (LMSA), embedded in a still larger scale synoptic rain area of lower rainrate. This structure is supported by radar and raingage observations of the type analyzed by Austin and Houze[1]; see Figure 4 for a schematic. While this structure is the supposed consequence of combined effects of vertical and horizontal motions, the precise dynamics of rainfall formation are not available. As a rule of thumb, the (possibly artificial) scales of these regions decrease by successive factors of 1/10 from the synoptic scale through LMSA, SMSA, and down to a cell, while the corresponding rainrates nearly double at each level until the scale of a cell where this rule generally breaks down; supercells are possible where the rainrate may be larger than the SMSA by several orders of magnitude.

One of the earliest studies of the spatial and temporal variability of rainfall was that of LeCam[51] based on spatial cluster point processes and random measures of the type also occurring in the study of the clustering of galaxies, earthquake aftershock sequences, population growth, etc. A substantial body of research has evolved over the past decade involving various problems associated with this approach to describing rainfall. Among these is a reasonably accurate computation of spatial/temporal correlation structure down to the scales of cells; see Zawadzki,[52] Waymire, Gupta, and Rodriguez-Iturbe,[53] Rodriguez-Iturbe, Cox, and Eagleson,[50] Phelan and Goodall,[54] Bell,[2] and Smith and Karr.[55] However, the lower scale high-intensity regions have not been adequately represented in this approach. From the point of view of the hydrologic and meteorologic applications the raincells are where the action is!

Considerations of scaling properties of rain data have led to random cascade models of the type described in section 2; see Lovejoy and Shertzer,[56] Tessier, Lovejoy, and Schertzer,[57] and Gupta and Waymire.[20,21] The basis for such models in the context of rainfall will be explained next.

CASCADE STRUCTURE OF SPATIAL RAINFALL. One of the fascinating symmetries observed in nature is that of self-similarity. The simplest form of this symmetry in a statistical context is that of simple scaling of the probability distributions. A random field $\{R(x)\}$ is said to be simple scaling if

$$\{R(\lambda x)\} = \{\lambda^\theta R(x)\} \tag{3.6}$$

where equality is in the sense of joint distributions. The *scaling exponent* θ is a real number parameter. One approach to testing this hypothesis is to compute moments of order h as a function of scale λ. One gets the log-log relation

$$\log\langle R^h(\lambda x)\rangle = h\theta \log \lambda + c_h. \tag{3.7}$$

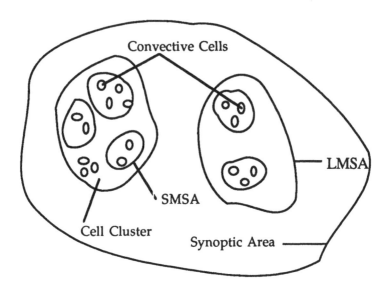

FIGURE 4 A schematic of spatial clustering observed in meso-scale rainfall (from Gupta and Waymire.[18]

In particular, simple scaling translates into two properties:

   i.   log-log linearity between a specified moment and length scale; and

   ii. a linear change in slope $s(h) = \theta h$ of the line as a function of moment order.

Familiar examples of simple scaling behavior include Gaussian white noise, fractional Brownian noise, increments of Brownian motion, and Levy stable processes.

    In the application of this test to spatial rainfall data, one finds a very interesting result. Namely, property (i) is preserved but the slope function $s(h)$ in (ii) is nonlinear; see Gupta and Waymire.[20] This is precisely the structure one may compute for a cascade distribution of rain. The essential computations are as follows. Let $\Delta_\lambda(i), i = 1, 2, \ldots$ denote a partition of a region X of d-dimensional space into cells at the length scale λ. Then

$$\log_b E \left[\sum_i \mu_\infty^h(\Delta_\lambda(i)) \right] = -d\chi_b(h)\log(\lambda) + \log E\mu_\infty^h(X), \qquad (3.8)$$

where

$$\chi_b(h) = \log_b EW^h - (h-1) \qquad (3.9)$$

and

$$\frac{\log_b \text{Prob}[\mu_\infty(\Delta_\lambda) > \lambda^{d\alpha}]}{\log \lambda} \to -d\chi_b^*(1 - \alpha), \quad \lambda \to 0 \tag{3.10}$$

where

$$\chi_b^*(a) = \sup_h [ah - \chi_b(h)] \tag{3.11}$$

is the Legendre transform of $\chi_b(h)$. Note from Eq. (3.8) that the structure function exponent for random cascades is given by $-d\chi_b(h)$.

In order to apply these formulae to sample realizations, one needs to be able to "drop the expectations." However because of long-range spatial correlations the cascade fields are nonergodic and one cannot simply replace expected values by spatial averages. Since data typically consists of a single spatial sample realization, this problem is quite important. Nonetheless it was partially solved by Holley and Waymire[24] for a large class of cascades, namely those with suitably bounded generators. In particular, the Beta model, for example, can be tested within this framework. In fact the Beta model provides an extremely simple yet accurate representation of the rainfall cluster and intermittency structure; see Over and Gupta.[58] Temporal evolutions of such models are also being considered by us. Higher order corrections to these models are also being pursued in extensions to cascades with statistically dependent generators (Waymire and Williams[37])

3.3 SCALING INVARIANCE IN REGIONAL FLOODS

INTRODUCTION AND A BIT OF HISTORY. An understanding of the spatial or regional structure of floods is a long-standing problem of great importance in hydrology. Kinnison and Kolby[29] introduced the index-flood assumption for regional flood analysis.floods, see also quantile regression approach It continues to be used extensively by the hydrologic research community. It states that annual floods at individual stations in a "homogeneous region," however defined, when scaled by their means have a common probability distribution or frequency curve.[6] Consequently, the scaled floods for all stations in the region have the same statistical moments, and this implies that the coefficient of variation (CV) of the floods also remains the same. In fact, the constancy of the CV is commonly used as a definition of regional homogeneity of floods.

During the late 1950, analysis of flood data from two regions in the midwest and the northeast U.S. showed that the index flood assumption did not hold.[5] This body of work within the U.S. Geological Survey (USGS) led to the development of the "quantile regression approach" (Benson[3]). It has been used extensively by the USGS since the mid-1960s. In this approach, each flood quantile, i.e., flows with specified probability of exceedance, is regressed against multiple basin characteristics, such as drainage area, mean basin slope etc., using multiple regression analysis. Although basin characteristics are used in the regional relations, there is no physical foundation for their inclusion or exclusion, and they are treated purely as statistical variables. Consequently, this line of investigation has remained data intensive and essentially statistical in nature. The second set of approaches to regional flood frequency have been built on the index-flood assumption as their underlying basis for development. The main emphasis has been on the development of statistically robust regional estimators of flood quantiles and the delineation of homogeneous regions. However, as stated above, the index flood assumption is generally belied by the extensive data provided by the USGS; see Gupta and Dawdy.[17] The third set of approaches to flood frequency analysis are based on the premise that by describing the basin response function and the rainfall input of a given frequency, the output for the flood peak of that frequency can be predicted. In this respect, they can be called "physically based" (Eagleson[8]; Sivapalan, Wood, and Beven[59]). The existing body of work on this topic has only focused on deriving flood frequency distribution at a fixed site rather than on their regional behavior. Therefore, connections and consistency between physically based approaches and regional approaches, either empirical or statistical, for the most part have remained unexplored; see Gupta et al.[16] for further details of these issues.

Even this brief historical overview reveals that these three sets of approaches to regional flood frequency analysis have remained largely disjoint. This state of affairs is not satisfactory, and illustrates a real need to develop a unified theoretical framework which includes data, the physics of the system and the statistics of the system. This furnishes the context within which the hypothesis of scaling invariance in floods and its implications for the development of a theory should be examined.[20,16,60]

SCALING THEORIES OF FLOODS. Consider a large drainage network, and assume that river discharges are being measured at every junction in this network. The variable of interest here is the maximum annual "instantaneous flow" at a gage, called the peak flow. The collection of peak flows or floods at each junction can be a viewed as a random field indexed by the junctions of the drainage network. The invariance of the joint probability distributions of peak flows under translation on this indexing set defines statistical homogeneity; see Gupta et al.,[16] and Waymire.[48] Since discharge at each junction necessarily depends on the size of the upstream subbasin contributing flows to this junction, drainage area should be included as a member of the indexing set. We define spatial homogeneity of floods to mean that flow at each junction is indexed by the drainage area A and nothing else. In other words, except for its dependence on drainage area A, the peak flow random field $Q(A)$ is translation invariant on the indexing set.

Both simple scaling and multiscaling theories described in section 3.2 have been tested on regional flood frequencies. In the context of floods, it is noteworthy that simple scaling is closely related to the widely used index flood assumption, and it predicts a constant CV.[16,60] Recent analyses of USGS flood data from New York, New Mexico, and Utah in the U.S. have suggested the hypothesis that snow-melt-generated floods exhibit simple scaling whereas rainfall-generated floods exhibit multiscaling.[17] Similarly, maximum likelihood tests of simple versus multiscaling for the Appalachia flood data rejected simple scaling over multiscaling; see Smith[60] for details.

Recall from our discussion in section 3.2 that both the simple and multiscaling theories require log-log linearity between the scale parameter A and the statistical moments $E[Q^j(A)]$. Let the slopes of these plots for each j be denoted by the function $s(j)$, which is the structure function exponent for floods. If the flood random field is simple scaling, then the function $s(j)$ is linear in j. On the other hand, if the random field is multiscaling, then $s(j)$ is nonlinear in j, and in particular is either concave or convex in j. This results in two distinct mathematical representations of a multiscaling process.[20] Concavity of $s(j)$ implies a decrease in spatial variability of floods with an increase in the spatial scale, while its convexity implies an increase in spatial variability with scale. In Figure 5 we show a schematic of how the CV varies with respect to scale. Since the CV is a measure of spatial variability of floods, Figure 5 exhibits the two distinct multiscaling representations.

Analysis of the Appalachia flood data first suggested that both of the multiscaling representations may be relevent for flood analysis.[16,60] The concavity of the structure function seems to hold only for basins larger than about 50 km^2;

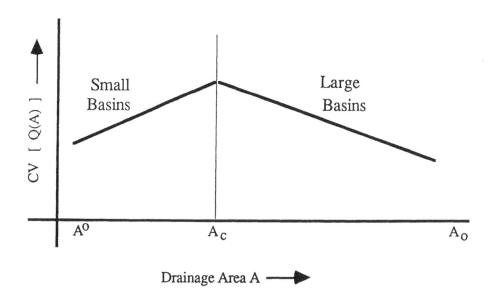

FIGURE 5   A schematic of the spatial variability of floods in small and large basins (from Gupta and Dawdy.[16]

see Smith[60] and Gupta et al.[16] Physically, the hypothesis of multiscaling in rainfall-generated floods suggests that they inherit this feature directly from the multiscaling structure of rainfall. By contrast, in smaller basins, the structure function exponent $s(j)$ seems to exhibit a convex behavior. Even though this conclusion is tentative, preliminary rainfall-runoff simulation experiments by Gupta and Dawdy[17] give support to this hypothesis on physical grounds. These results suggest that spatial rainfall is the dominant factor underlying the scaling structure of floods in large basins, where as the geometry of channel network determines the structure function of floods in small basins. In this sense the multiscaling theory provides the first underpinnings of a framework which can combine data, hydrology, and the statistics of floods. We will discuss this topic of synthesis next.

4. CONCLUSIONS AND FUTURE DIRECTIONS: TOWARDS A SYNTHESIS

A major question in the development of a theory of floods is: What have we learned from the connections described in Section 3 that can help to simplify the problem without losing track of its physical relevance? Let us enumerate the following points in this regard. First, one can simplify the problem physically by initially ignoring the various runoff-generating mechanisms on the hill slopes. Sivapalan et al.[59] have investigated the dependence of the flood frequency function on eight dimensionless parameters called similarity parameters. These parameters incorporate different physical processes related to rainfall and rainfall-runoff transformations. This work showed that the three most significant factors influencing flood frequency at a station, in the decreasing order of importance, are the drainage area, mean precipitation, and the mean saturated hydraulic conductivity. This finding has been used by Gupta and Dawdy[16] to suggest a physical interpretation of two important parameters appearing in a multiscaling representation of floods for large basins. Second, data analyses of floods suggest that precipitation is the key physical mechanism determining the scaling structure of floods in large basins. Third, preliminary rainfall-runoff simulation suggest that the geometry of channel network determines the structure function of floods in small basins, but its influence on the structure function of floods in large basins remains unclear.

The three results summarized above point to an important first step in developing a scaling theory of floods. This step consists of understanding how a channel network partitions a drainage basin into smaller subbasins. This partitioning is necessary in order to distribute rainfall intensities spatially within a cascade framework. Once we know how rainfall can be distributed spatially on a basin, this water can be routed through the channel network using a simple routing equation, and the peak flows can be sampled over successively larger subbasins to investigate their spatial scaling structure. Both analytical and simulation studies would be necessary for this purpose.

Partitioning of a drainage basin appropriately by a channel network is related to the general problem of how a network develops under spatial constraints. As already discussed in section 3.1, it has long been recognized Some recent work related to this problem is that of Troutman and Karlinger[61] and, on the temporal evolution of river networks, that of Rodriguez-Iturbe, Marani, Rigon, and Rinaldo,[62] Leheny and Nagel,[63] Kramer and Marder,[64] and Peckham.[41] However, the general problem of how a channel network, e.g., an SST, divides a drainage basin into successively smaller subbasins remains unsolved. To illustrate this issue further, consider the Peano tree described in section 3.1. The mass distribution at each stage of construction of the width function corresponds to the proportion of drainage area that flows into the corresponding links. These drainage areas are shown with a dotted line in Figure 2. In other examples of SSTs, e.g., the Mandelbrot and Viscek,[46] this correspondence between drainage areas and the cascade masses is lost. Our ongoing work shows that even though the width functions for these other SSTs map them onto multinomial cascades, the cascade masses for these trees do not correspond to drainage areas; see Gupta et al.[17] Therefore an understanding of how an SST subdivides a basin is an important first step towards a synthesis of rainfall and landforms in developing a space-time theory of river run-off and floods.

ACKNOWLEDGEMENTS

This resesarch was supported by grants from NSF and NASA.

REFERENCES

1. Austin, P. M., and R. A. Houze. "Analysis of the Structure of Precipitation Patterns in New England." *J. Appl. Met.* **11** (1972): 926–935.
2. Bell, T. L. "A Space-Time Stochastic Model of Rainfall for Satellite Remote Sensing Studies." *J. Geophys. Res.* **92(D8)** (1987): 9631–9643.
3. Benson, M. A. "Factors Influencing the Occurrence of Floods in a Humid Region of Diverse Terrain." U.S. Geological Survey Water Supply Professional Paper 1580B, 1962, p. 62.
4. Chayes, J, L. Chayes, and R. Durrett. "Connectivity of Mandelbrot's Percolation Process." *Prob. Theor. Related Fields* **77** (1988): 307–324.
5. Dawdy, D. R. "Variation of Flood Ratios with Size of Drainage Area." U.S. Geological Survey Research, 424-C, Paper C36, Reston, VA, 1961.

6. Dalrymple, T. "Flood-Frequency Analyses." U.S. Geological Survey Water Supply Paper No. 1543-A, 1960, p. 80.

7. Dubrulle, B. "Intermittency in Fully Developed Turbulence: Log-Poisson Statistics and Generalized Scale Invariance." *Phys. Rev. Lett.* **73(7)** (1994): 959–962.

8. Eagleson, P. S. "Dynamics of Flood Frequency." *Water Resour. Res.* **8(4)** (1972): 878–898.

9. Evertsz, C. J. G., and B. B. Mandelbrot. In *Chaos and Fractals: New Frontiers of Science, Appendix B on Multifractal Measures*, edited by H. O. Peitgen, H. Jergens, and D. Saupe. New York: Springer-Verlag, 1992.

10. Falconer, K., ed. *Fractal Geometry: Mathematical Foundations and Applications.* New York: John Wiley and Sons, 1990.

11. Feder, J. *Fractals.* New York: Plenum Press, 1988.

12. Feigenbaum, M., M. H. Jensen, and I. Procaccia. "Time Ordering and Thermodynamics of Strange Sets: Theory and Experimental Tests." *Phys. Rev. Lett.* **57(13)** (1986): 1503–1506.

13. Gupta, V. K., and D. R. Dawdy. "Regional Analysis of Flood Peaks: Multiscaling Theory and Its Physical Basis." In *Advances in Distributed Hydrology*, edited by R. Rosso, A. Peano, I. Becchi, and G. A. Bemporad, 149–168. Highland Ranch, CO: Water Resources Publications, 1994.

14. Gupta, V. K., and D. R. Dawdy. "Physical Interpretations of Regional Variations in the Scaling Exponents of Flood Quantiles." *Hydrological Processes*, 1995.

15. Gupta, V. K., and O. Mesa. "Runoff Generation and Hydrologic Response via Channel Network Geomorphology: Recent Progress and Open Problems." *J. Hydrol.* **102** (1988): 3–28.

16. Gupta, V. K., O. Mesa, and D. R. Dawdy. "Multiscaling Theory of Flood Peaks: Regional Quantile Analysis." *Water Resour. Res.* **30(12)** (1994): 3405–3421.

17. Gupta, V. K., S. Peckham, S. Veitzer, and E. Waymire. "Width Functions of Self-Similar Trees; Isomorphisms to Multinomial Cascades." 1995 (in preparation).

18. Gupta, V. K., and E. Waymire. "A Stochastic Kinematic Study of Subsynoptic Space-Time Rainfall." *Water Resour. Res.* **15(3)** (1979): 637–644.

19. Gupta V. K., and E. Waymire. "Statistical Self-Similarity in River Networks Parametrized by Elevation." *Water Resour. Res.* **25(3)** (1989): 463–476.

20. Gupta, V. K., and E. Waymire. "Multiscaling Properties of Spatial Rainfall and River Flow Distributions." *J. Geophys. Res.* **95(D3)** (1990): 1999–2009.

21. Gupta, V. K., and E. Waymire. "A Statistical Analysis of Mesoscale Rainfall as a Random Cascade." *J. Appl. Met.* **12(2)** (1993): 251–267.

22. Gupta, V. K., E. Waymire, and I. Rodriguez-Iturbe. "On Scales, Gravity and Network Structure in Basin Runoff." In *Scale Problems in Hydrology*, edited by V. K. Gupta, I. Rodriguez-Iturbe, and E. Wood. Dordrecht, The Netherlands: D. Reidel, 1986.

23. Hack, J. T. "Studies of Longtudinal Stream Profiles in Virginia and Maryland." U.S. Geological Survey Prof. Paper 294-B, 505-B, 1957.
24. Holley, R., and E. Waymire. "Multifractal Dimensions and Scaling Exponents for Strongly Bounded Random Cascades." *Ann. Appl. Prob.* **2(4)** (1992): 819–845.
25. Jarvis, R. S., and M. J. Woldenberg, eds. *River Networks.* Benchmark Papers in Geology, Vol 80. Stroudsburg, PA: Hutchinson Ross, 1984.
26. Katz, R. "Probabilistic Models." In *Probability, Statistics, and Decision Making in Atmospheric Sciences*, edited by A. Murphy and R. Katz. Boulder, CO: Westview Press, 1985.
27. Kahane, J. P. "Positive Martingales and Random Measures." *Chinese Ann. Math.* **8b** (1987): 1–12.
28. Karlinger, M., and B. Troutman. "A Random Spatial Network Model Based on Elementary Postulates." *Water Resour. Res.* **25** (1989): 793–798.
29. Kinnison and Kolby. "Flood Formulas Based on Drainage Basin Characteristics." *ASCE Trans.* **110** (1945): 849–904.
30. Kirby, M. J. "Tests of Random Model, and Its Applications to Basin Hydrology." *Earth Surface Processes* **1** (1976): 197–212.
31. Kolmogorov, A. N. "Local Structure of Turbulence in an Incompressible Liquid for Very Large Reynolds Numbers." *Comptes Rendus (Doklady) de l'academie des sciences de l'URSS* **30** (1941): 301–305.
32. Kolmogorov, A. N. "A Refinement of Previous Hypotheses Concerning the Local Structure of Turbulence in a Viscous Inhomogeneous Fluid at High Reynolds Number." *J. Fluid Mech.* **13** (1962): 82–85.
33. Kramer, S., and M. Marder. "Evolution of River Networks." *Phys. Rev. Lett.* **68(2)** (1992): 205–208.
34. LeCam, L. "A Stochastic Description of Precipitation." In *Fourth Berkeley Symposium on Mathematical Statistics, and Probability*, Vol. 3, 165–186. Berkeley, CA: University of California Press, 1961.
35. Leheny, R., and S. Nagel. "Model for the Evolution of River Networks." *Phys. Rev. Lett.* **71(9)** (1993): 1470–1473.
36. Leopold, L. B., M. G. Wolman, and J. P. Miller. *Fluvial Processes in Geomorphology.* San Francisco, CA: W. H. Freeman, 1964.
37. Lovejoy, S., and B. B. Mandelbrot. "Fractal Properties of Rain and a Fractal Model." *Tellus* **37A** (1985): 209–232.
38. Lovejoy, S., and D. Schertzer. "Generalized Scale Invariance in the Atmosphere and Fractal Models of Rain." *Water Resour. Res.* **21(8)** (1985): 1233–1250.
39. Lovejoy, S., and D. Schertzer. "Multifractals, Universality Classes, and Satellite and Radar Measurements of Cloud and Rain Fields." *J. Geophys. Res.* **95(D3)** (1990): 2021–2031.
40. Mandelbrot, B. B. "Intermittent Turbulence in Self-Similar Cascades: Divergence of High Moments and Dimension of the Carrier." *J. Fluid Mech.* **62** (1974): 331–358.

41. Mandelbrot, B. B. *The Fractal Geometry of Nature*. San Francisco, CA: Freeman, 1982.

42. Mandelbrot, B. B., and T. Viscek. "Directed Recursive Models for Fractal Growth." *J. Phys. A* **22** (1989): L377–L383.

43. Maxwell, M. "Asymptotic Enumeration for Combinatorial Structures." Ph.D. Dissertation, Oregon State University, Corvallis, OR, 1994.

44. Marani A., R. Rigon, and A. Rinaldo. "A Note on Fractal Channel Networks." *Water Resour. Res.* **27(12)** (1991): 3041–3049.

45. Over, T., and V. Gupta. "Statistical Analysis of Meso-Scale Rainfall: Dependence of a Random Cascade Generator on Large-Scale Forcing." *J. Appl. Met.* **33(12)** (1994): 1526–1542.

46. Peckham, S. "New Results for Self-Similar Trees with Applications to River Networks." *Water Resour. Res.* **31(4)** (1995): 1023–1029.

47. Peckham, S. "Self-Similarity in the Three-Dimensional Geometry and Dynamics of Large River Basins." Ph.D. dissertation, University of Colorado, Boulder, CO, 1995.

48. Phelan, M. J., and C. R. Goodall. "An Assessment of a Generalized Waymire-Gupta-Rodriquez-Iturbe Model for GARP Atlantic Tropical Experimental Rainfall." *J. Geophys. Res.* **95(D6)** (1990): 7603–7615.

49. Riddell, R. J., and G. E. Uhlenbeck. "On the Theory of Virial Development of the Equation of State of Monoatomic Gases." *J. Chem. Phys.* **21** (1953): 2056–2064.

50. Rodriguez-Iturbe, I. "Scale of Fluctuation of Rainfall Models." *Water Resour. Res.* **22** (1986): 155–375.

51. Rodriguez-Iturbe, I., D. R. Cox, and P. S. Eagleson. "Spatial Modeling of Total Storm Rainfall." *Proc. Roy. Soc. Lond., Ser. A.* **403** (1986): 27–50.

52. Rodriguez-Iturbe, I., M. Marani, R. Rigon, and A. Rinaldo. "Self-Organized River Basin Landscapes: Fractal and Multifractal Characteristics." *Water Resour. Res.* **30(12)** (1994): 3531–3539.

53. Smart, J. S. "Channel Networks." *Adv. Hydro. Sci.* **8** (1972): 305–346.

54. She, Z. S., and E. Leveque. "Universal Scaling Laws in Fully Developed Turbulence." *Phys. Rev. Lett.* **72** (1994): 336.

55. She, Z. S., and E. C. Waymire. "Quantized Energy Cascade and Log-Poisson Statistics in Fully Developed Turbulence." *Phys. Rev. Lett.* **74(2)** (1995): 262–265.

56. Shreve, R. L. "Infinite Topologically Random Channel Networks." *J. Geol.* **75** (1967): 178–186.

57. Sivapalan, M., E. F. Wood, and K. J. Beven. "On Hydrologic Similarity, 3, Adimensionless Flood Frequency Model Using a Generalized Geomorphic Unit Hydrograph and Partial Area Runoff Generation." *Water Resour. Res.* **26(1)** (1990): 43–58.

58. Smith, J. "Representation of Basin Scale in Flood Peak Distributions." *Water Resour. Res.* **28(11)** (1992): 2993–2999.

59. Smith, J. A., and A. F. Karr. "Parameter Estimation for a Model of Space-Time Rainfall." *Water Resour. Res.* **21(8)** (1985): 1251–1257.
60. Tessier, Y., S. Lovejoy, and D. Schertzer. "Universal Multifractals: Theory and Observations for Rain and Clouds." *J. Appl. Met.* **32(2)** (1993): 223–250.
61. Tokunaga, E. "Ordering of Divide Segments and Law of Divide Segment Numbers." *Trans. Japanese Geomorph. Union* **5(2)** (1984): 71–77. (In English).
62. Troutman, B. M., and M. Karlinger. "On the Expected Width Function of Topologically Random Channel Networks." *J. Appl. Prob.* **22** (1984): 836–849.
63. Troutman, B. M., and M. R. Karlinger. "Gibbs' Distribution on Drainage Networks." *Water Resour. Res.* **28(2)** (1992): 563–577.
64. Waymire, E. C. "Scaling Limits and Self-Similarity in Precipitation Fields." *Water Resour. Res.* **21(8)** (1985): 1271–1281.
65. Waymire, E. C. *On Network Structure Function Computations.* Proceedings of the Institute for Mathematics and Its Applications. New York: Springer-Verlag, 1991.
66. Waymire, E. C., V. K. Gupta, and I. Rodriguez-Iturbe. "A Spectral Theory of Rainfall Intensity at the Meso-Scale." *Water Resour. Res.* **20(10)** (1984): 1453–1465.
67. Waymire, E. C., and S. Williams. "A General Decomposition Theory for Random Cascades." *Bull. Amer. Math. Soc.* **31(2)** (1994): 216–222.
68. Waymire, E., and W. Williams. "Multiplicative Cascades: Dimension Spectra and Dependence." *J. Fourier Analysis and Appl.* (1995): In press.
69. Wood, E. F., M. Sivapalan, and K. J. Beven. "Similarity and Scale in Catchment Storm Response." *Rev. Geophys.* **28(1)** (1990): 1–18.
70. Zawadzki, I. I. "Statistical Properties of Precipitation Patterns." *J. Appl. Met.* **12** (1973): 459–472.
71. Zhang, Q. "Fluctuations in the Width Function for the Random Model." Ph.D. Thesis, Oregon State University, Corvallis, OR, 1995.

Brent M. Troutman
U.S. Geological Survey, Denver Federal Center, Box 25046, MS 418, Denver, CO 80225

Inference for a Channel Network Model and Implications for Flood Scaling

We discuss statistical inference for a two-parameter (β_0 and β_1) Gibbsian probability model for channel networks defined on a two-dimensional square lattice. The probability of a tree s is proportional to $\exp[-\beta_0 H(s, \beta_1)]$, where $H(s, \beta_1)$ is taken to be a summation over lattice points ν in the basin of $A(\nu, s)^{\beta_1}$, letting $A(\nu, s)$ be the number of points upstream from point ν. $A(\nu, s)$ may be interpreted as drainage area associated with ν. Procedures are discussed for estimating the parameters given an actual network obtained from digital elevation data, for using bootstrapping to obtain confidence intervals for these estimates, and for testing model goodness-of-fit. The connection between β_1 and scaling exponents in discharge/slope/ drainage area relationships is discussed. The exponent in the power law relating discharge to drainage area is of particular relevance in flood hazard studies. The statistical inference procedures are applied to a set of data from Willow Creek in Montana, for which estimates of the parameter β_1 tend to be centered around 0.75. It is shown that there is a close connection between the estimates of β_1 and the estimated exponent γ in the scaling law relating slope to drainage area. We also argue that the statistical procedures developed here will be useful for rigorously testing optimal channel network models proposed by other researchers.

INTRODUCTION

Hydrologists have studied extensively the problem of flood prediction. Particularly challenging is the problem of flood hazard estimation at locations that are ungaged or that have only short records. Regional regression, in which flood properties are regressed against basin physiographic and climatic characteristics, is now widely used as a tool for transferring information from gaged to ungaged catchments. Drainage area is the physiographic characteristic that most often shows up as a significant predictor of basin hydrologic response. In addition to its usefulness in empirical hydrologic modeling, however, drainage area is also a logical choice for the size parameter in studies of scaling properties of hydrologic processes. Forms of scaling invariance have shown up in a number of ways in drainage basin morphometry and in analysis of hydrologic fluxes such as precipitation. Understanding these scaling properties better is not only important from the scientific point of view, but also holds the promise of better technologies for flood prediction.

In the next section, we discuss a few of the scaling laws, including the power law relating water discharge (Q) to drainage area (A), that have been investigated by hydrologists. Of particular interest to us is a quantity U which is proportional to Q multiplied by slope S. This quantity has been studied in the context of rate of potential energy expenditure, or "stream power," as QS is easily shown to be proportional to time rate of change of mass times change in elevation. Recent studies (e.g., Rinaldo et al.[11]) have investigated the question of whether channel networks tend to minimize a spatially integrated value of U, which at each point is taken to scale as $A^{0.5}$. We point out below a few of the factors rendering this exponent of 0.5 uncertain, thus motivating our approach of defining a two-parameter Gibbsian probability model on channel networks and developing procedures for statistical inference. One parameter of this model, denoted by β_1, is the exponent of A, and the second parameter, β_0, is the coefficient of the spatially summed U and is analogous to inverse temperature in the canonical distribution. We illustrate how these procedures provide information on the appropriate value of the exponent of A and also on how we may rigorously evaluate claims that actual networks minimize an energy functional.

SCALING LAWS

To illustrate the approach and to show how scaling laws enter, let us discretize a drainage network and express energy dissipation in the ith segment of the network as

$$U_i = Q_i S_i L_i \tag{1}$$

where Q_i is discharge from the ith segment, L_i is length of the ith segment, and S_i is the slope of the ith segment. We may think of a "segment" here as being either a link or, if the network is defined on a square lattice, a portion of a channel connecting two adjacent grid points. We shall use the latter interpretation in this paper. In this situation, L_i may be assumed to be constant, and will be taken to be 1 throughout. Also, because discharge is time-varying, we may remove time dependence from the problem by assuming that Q_i stands for discharge at some fixed time at which certain representative flow conditions (such as mean flow or bankfull flow) hold, or alternatively that Q_i stands for some functional of the joint flow distributions over a fixed period of time; for example, Q_i may be the 50-year flood. Finally, we may also associate a drainage area, say A_i, with the ith segment.

Scaling arguments typically take the following form. If it is assumed that

$$S \sim Q^\alpha, \tag{2}$$

then U_i scales as

$$U_i \sim Q_i^{1+\alpha}. \tag{3}$$

If it is further assumed that

$$Q_i \sim A^\theta, \tag{4}$$

then we have

$$U_i \sim A_i^{(1+\alpha)\theta}. \tag{5}$$

Total (spatially summed) energy dissipation in the entire drainage basin, H, then obeys

$$H = \sum_i U_i \sim \sum_i A_i^{(1+\alpha)\theta} \tag{6}$$

where the summation is over all segments in the basin. If it is assumed that $\alpha = -0.5$ (for instance, Leopold et al.,[9] page 244, report a value of -0.49 for mean annual flow conditions in the midwestern United States) and that discharge Q is proportional to area A, i.e., $\theta = 1$, then Eq. (6) becomes

$$H \sim \sum_i A_i^{0.5}. \tag{7}$$

Although their argument is more detailed than what we have presented here, Rodriguez-Iturbe et al.[14] obtain an expression identical to Eq. (7) using optimal energy expenditure principles. They also present evidence that actual networks tend to minimize Eq. (7) (see also Rinaldo et al.[11,12]).

We now wish to bring to light some issues that illustrate uncertainties associated with the exponent of A in Eq. (5). First, the exponents α in Eq. (2) and θ in Eq. (4) are subject to error. One critical question seems to be: What is the appropriate definition and interpretation of "energy expenditure" in view of the temporal variability of discharge, and, correspondingly, what is the best way to estimate the scaling exponents given this temporal variability? We stated above that we wish to remove the temporal dependence by taking Q to be some representative flow amount, but it is not clear what "representative" should mean here. If Q is, say, mean annual discharge, then a value of $\theta = 1$ in Eq. (4) may be appropriate. But if energy expenditure ideas are going to be applied to look at, for example, channel network configurations, which may very well be largely influenced by more extreme discharges, then it may be more appropriate to take Q in Eq. (1) to correspond, say, to a quantile of the probability distribution of annual peak flow. In this case empirical studies have shown that the value of θ will change with return period, and will tend to be less than 1 (see, for example, Gupta et al.[4] and Smith[16]). Similarly, the relationship between slope and discharge has been investigated empirically by a number of investigators, and a wide range of values of α have been reported, depending on region and conditions (see Leopold et al.,[9] Leopold and Wolman,[8] Leopold and Maddock,[7] Flint,[3] and Wolman[22]). Sun et al.[17] investigate the geometry of channel networks obtained by minimizing Eq. (6) for different values of α and with $\theta = 1$.

In addition to uncertainties in the scaling exponents in relations involving Q, S, and A, there are some more subtle issues regarding the legitimacy of the algebraic manipulations that are performed in arriving at an expression like Eq. (6). Such manipulations work when dealing with exact (or almost exact) relationships among variables, but the relationships between Q, S, and A in the present problem are subject to a good deal of uncertainty and scatter, so we need to be more precise in stating exactly how these relationships are to be interpreted. Let us consider the trivariate spatial process (Q, S, A). For now we may assume that these three variables are nonrandom at a given grid point, but we introduce randomness by sampling the grid points in such a way that the n points in a basin are equally likely (uniform distribution). Taking $U = QS$, we see that H is simply $nE(U)$, where expectation here stands for a spatial average obtained by sampling under our assumed uniform distribution on grid points. Given scaling relations involving the variables Q, S, and A, we now ask what can be deduced about the scaling of U as a function of A.

We shall assume the scaling relations in Eqs. (2) and (4) to be defined in terms of mean values. Thus, we write Eq. (2) as

$$E(S|Q) = aQ^\alpha. \tag{8}$$

The exponent α in Eq. (8) is typically estimated by performing a linear regression of $\log(S)$ versus $\log(Q)$, but the scaling law (8) does not in general imply a log-linear relationship between $\log(S)$ and $\log(Q)$, particularly if the coefficient of variation

$CV(S|Q)$ is not constant, so there may be serious difficulties associated with an ordinary log-linear least-squares fitting procedure. Evidence for the nonconstancy of the CV for S as a function of A is presented by Tarboton et al.[18]; and it is expected that such nonconstancy for S as a function of Q will also hold in general. (Scaling properties of the slope-area relationship are investigated also by Gupta and Waymire.[5]) Furthermore, if we write Eq. (4) more precisely as

$$E(Q|A) = bA^\theta, \tag{9}$$

then it is readily seen that

$$E(U) = aE(Q^{1+\alpha}) \neq abE(A^{(1+\alpha)\theta}) \tag{10}$$

except under very special conditions. Hence, when scaling laws are interpreted in this manner, it is seen that the result obtained in Eq. (6) by loose arguments is not correct. In general, properties of the distribution of Q given A other than the first moment in Eq. (9) will influence the scaling in Eq. (10). Variability of Q for a given area A is related to the numerous physiographic and climatic factors influencing runoff; properties of this conditional distribution have been investigated in the context of regional flood frequency regression (e.g., Lichty and Karlinger[10] and Tasker and Stedinger[19]).

Another way to view this problem is as follows. Rather than the slope-discharge scaling in Eq. (8), let us look at slope-area scaling; assume that slope and area are related by

$$E(S|A) = cA^\gamma. \tag{11}$$

One might be tempted to conclude that Eqs. (9) and (11) imply a scaling exponent of $\gamma + \theta$ in $E(U)$, but in fact

$$E(U) \neq cbE(A^{\gamma+\theta}), \tag{12}$$

again except under special conditions (for example, conditional independence of Q and S, which most probably does not hold). In general, one must know something about the joint distribution of Q and S given A to properly determine the scaling exponent for A.

The conclusion, then, is that it is in general not possible to obtain the appropriate exponent of A in computing $E(U)$, and hence in computing H, given only scaling relationships such as those in Eqs. (8), (9), and (11). More detailed knowledge (beyond conditional first moments) of the trivariate process (Q, S, A) is needed.

Finally, given that one has decided on an appropriate energy functional such as Eq. (7), there remains the problem of determining whether actual networks really minimize H. There are several problems. First, there are a large number of configurations that will yield an optimal value of H, so we are really asking whether an actual network belongs to this class. Secondly, if a numerical optimization algorithm such as simulated annealing, using Eq. (7), say, with a fixed value of the exponent of A as the objective function, is applied with the actual network as the starting point, it will almost always be the case that lower energy networks will result. We may then legitimately ask how far an actual H may deviate from the optimal value for the actual network to still be considered an "optimal channel network."

These points motivate the features of the approach that we take here. First of all, it is clear that there are many uncertainties in determining the appropriate exponent of A in the energy functional H; these uncertainties are connected with the hydraulics arguments, interpretations, variability in the scaling relations among Q, S, and A, and data analysis procedures. It is our belief that a more defensible approach at this point is to first solve a simpler problem which can be analyzed strictly in terms of geometry, and then, once this problem is understood, to take the steps necessary to establish connections with the movement of water. Thus, our formulation will begin with an expression like Eq. (6) involving only area A, and analysis will be done using only geometric information that is a function of digitized elevation data.

We apply a statistical procedure known as maximum likelihood estimation to networks obtained from digital elevation data to obtain an estimate of the exponent of area, to be denoted by β_1, and to quantify how much uncertainty there is in the estimated value. There is a second parameter, β_0, in our model, which will tend to be infinitely large for optimal channel networks. Thus, by estimating these two parameters jointly, our statistical inference procedures provide us with a direct means of testing whether real networks really do tend to minimize an energy functional. Finally, statistical analysis allows us to objectively define and test "goodness-of-fit"; that is, we can quantify how far a real network is from "optimal" and whether this deviation is significant.

METHODOLOGY

SPATIAL NETWORK MODEL

We present here a brief description of the probability model to be applied to channel networks; a more detailed discussion may be found in a paper by Troutman and Karlinger.[21] We begin with a square grid over a given drainage basin B; let $V = V(B)$ denote the lattice points in B, set $n = |V|$, the number of points in the set V, and designate one of the points in V as the outlet. Our sample space $S = S(B)$ is the set of all rooted (at the fixed outlet) spanning trees of B. We shall define a Gibbsian-type model on this set of spanning trees by taking the probability of a tree t to be given by the two-parameter probability distribution

$$P_\beta[t] = Z^{-1} e^{-\beta_0 H(t,\beta_1)} \tag{13}$$

for $t \varepsilon S$, where $Z = Z(\beta)$ is the partition function, $\beta = (\beta_0, \beta_1)$ is a two-dimensional parameter, and the Hamiltonian $H(t, \beta_1)$ is a function of the tree t. The parameter space here is $\beta_0 \geq 0$ and $\beta_1 \geq 0$. We shall take H to be of the form

$$H(t, \beta_1) = \sum_{\nu} A(\nu, t)^{\beta_1}, \tag{14}$$

where, for the given tree t, $A(\nu, t)$ is the number of points (not including ν itself) upstream of the point ν in B. $A(\nu, t)$ may be interpreted as the drainage area associated with point ν, so H is a spatial summation over all the points in the basin B of area to the β_1 power.

There are several special cases of particular interest, including those mentioned in the section above. When $\beta_0 = 0$, the dependence of the probability of a tree on H is eliminated, and all trees become equally likely, yielding a uniform model on spanning trees. This model, analogous to the random topology model for trees that are not spatially embedded (see, for example, Shreve[15] and Abrahams[1]), was investigated by Karlinger and Troutman[6] and subjected to further statistical tests in a later paper by Troutman and Karlinger[20] that showed it to be unrealistic. Generally speaking, trees under this distribution tend to have channels that are too sinuous.

When $\beta_1 = 1$, it may be shown that

$$H(t, 1) = \sum_{\nu} A(\nu, t) = \sum_{\nu} d(\nu) \tag{15}$$

where $d(\nu) =$ the flow distance of point ν to the outlet. Thus, H is the spatial sum of flow distances, and therefore directly measures channel sinuosity. With β_1 fixed at 1, our model then becomes a one-parameter (β_0) model. This form of the model (with β_1 fixed at 1) was studied in detail by Troutman and Karlinger.[20]

When β_1 is set to 0.5, $H(t, 0.5)$ is essentially the expression for H in Eq. (7). Letting the inverse temperature β_0 tend to infinity with the exponent β_1 held at 0.5 yields a probability distribution for which the set of all trees minimizing $H(t, 0.5)$ are equally likely. These are the so-called "optimal channel networks" of Rodriguez-Iturbe et al.[14]

POINT ESTIMATION OF β

Let us look then more closely at the estimation problem. The situation is that we are given an observed tree, say t_0, and want to infer what β_0 and β_1 are. We have used the method of maximum likelihood; this method simply chooses parameter estimates so as to maximize the probability distribution with respect to the parameters with the tree t in Eq. (13) fixed at t_0. We may equivalently minimize the negative logarithm of the distribution, which is given by:

$$L(\beta, t_0) = -\log P_\beta[t_0] = \log Z(\beta) + \beta_0 H(t_0, \beta_1); \tag{16}$$

to minimize this we differentiate with respect to β_0 and β_1, and set these equations equal to 0. The final estimation equations are

$$E_\beta[H(T, \beta_1)] = H(t_0, \beta_1) \tag{17}$$

and

$$E_\beta[G(T, \beta_1)] = G(t_0, \beta_1), \tag{18}$$

where G is given by

$$G(t, \beta_1) = \sum_\nu (A(\nu, t))^{\beta_1} \log A(\nu, t). \tag{19}$$

In Eqs. (17) and (18), T denotes a random Gibbsian tree, and E_β denotes the mean value, or expectation, with respect to the Gibbsian model with parameters β. We have two equations in two unknowns which we solve for the desired parameter estimates. Thus, all the estimation procedure does is to pick β so as to ensure agreement between the observed values of two tree functions, H and G, and the corresponding mean values under the probability model.

The main problem with estimating the parameters is the fact that the expectations on the left-hand side of Eqs. (17) and (18) are not known explicitly. We must therefore resort to a Monte Carlo approach for obtaining estimates of these. The most desirable procedure would be of course to generate independent realizations of trees from P_β, but this is not feasible because (1) the sample space S is too large, and (2) the normalizing function Z is not known explicitly. Thus we make use of a Markov chain simulation procedure.

The goal is this: given a set of parameters, say β^*, we want to generate trees $T_1, T_2, \ldots$ according to a Markov chain such that the stationary distribution is P_{β^*}. The algorithm is this:

1. define an initial tree T_0;
2. for the given tree, randomly select (a) a point $\nu \varepsilon V$, with all n points equally likely, and (b) a new flow direction from ν, with all directions equally likely;
3. check whether the resulting structure is a legitimate tree (i.e., has no loops); if not, then take the new tree to be identical to the given tree and return to step 2, and otherwise go to step 4; and
4. draw X from a uniform distribution on the interval $(0, 1)$. If $X < e^{-\beta_0^* \Delta H}$, where ΔH is H for the new tree minus H for the given tree, then accept the new flow direction at ν and return to step 2; if not, then again take the given tree to be unchanged and return to step 2.

It may be shown that this algorithm obeys the reversibility (detailed balance) condition and the ergodicity condition required for obtaining the desired stationary distribution.

To estimate expectations, then, we use the result (see, for example, Aldous[2]) which shows that the weighted sample mean of a functional h of N trees

$T_1, T_2, \ldots, T_N$ generated by this Markov chain approach converges almost surely to the desired true mean:

$$\frac{\frac{1}{N}\sum_{i=1}^{N} w_i h(T_i)}{\frac{1}{N}\sum_{i=1}^{N} w_i} \longrightarrow E_\beta[h(T)] \tag{20}$$

almost surely, where

$$w_i = \exp[-\beta_0 H(T_i, \beta_1) + \beta_0^* H(T_i, \beta_1^*)]. \tag{21}$$

The problem is that this estimate may not be very good if β is not near to β^* and N is not large enough.

Thus, parameter estimation is done by an iterative procedure, yielding $\beta_0^*, \beta_1^*, \beta_2^*, \ldots$, converging to the solution of the estimation equations. The steps are: given β_j^*, generate trees and estimate expectations on the left-hand side of the estimation equations (17) and (18), and then re-solve the equations to obtain β_{j+1}^*. For details, the reader is referred to Troutman and Karlinger.[21]

UNCERTAINTY IN ESTIMATION OF β

In addition to obtaining point estimates of the parameters β, we would also like to have an idea of the uncertainty in these estimates. There are generally two sources of uncertainty: sampling uncertainty, and uncertainty due to errors in estimating $E_\beta[H]$ and $E_\beta[G]$ via Markov chain simulation. Sampling uncertainty comes from the fact that, for a given "true" set of parameters β, there are many trees t_0 that might be sampled, and this causes variability in the parameter estimates. This source of error may be estimated by a procedure known as bootstrapping, which, in general terms, is this:

1. given the estimate b of β, we then generate a large number, say I, of trees from the Gibbsian distribution P_b;
2. these I trees are used in turn as "observed" trees to obtain I bootstrap estimates of b (which is now taken to be the "true" parameter set), say $b(i)$, $i = 1, 2, \ldots, I$; thus, the generated trees are used one-at-a-time for t_0 in Eqs. (17) and (18); and
3. the resulting empirical distribution of the $b(i)$ may then be used to estimate the sampling properties of b. We have used the simplest procedure, known as the "percentile method."

Limits (L_j, U_j) for a 95% confidence interval for b_j are the 2.5% and 97.5% quantiles of the sample distribution of the $b_j(i)$.

Generally, the uncertainty due to the errors in estimating the expectations on the left-hand side of Eqs. (17) and (18) may be looked at by performing the parameter estimation with different seeds for the Monte Carlo Markov chain procedure and looking at the variability of parameter estimates. This source of uncertainty is generally small compared to sampling uncertainty.

GOODNESS-OF-FIT

In addition to the question of parametric inference, we need to address the question of goodness-of-fit. That is, we know how to obtain parameters that best match an actual network. But given these optimal parameters, we want to know how closely the model fits (i.e., mimics properties of) the real world. To address this question we first have to ask: What are the best ways to quantify discrepancies between model-generated trees and real trees? There are obviously many properties of trees that are important, but because the area distribution seems to be emerging as an important property in drainage basin geometry, we have developed tests based on it.

With each grid point we may associate an upstream drainage area; if there are n grid points, then we obtain the empirical distribution of these n areas. One statistic that has proven to be useful in measuring agreement between actual and model-generated area distributions is the Kolmogorov distance. In particular, if $F(., t)$ denotes the cumulative distribution function of area for a tree t, we define a test statistic D as

$$D = \sup |F(a, t_0) - E_b F(a, T)| \qquad (22)$$

where the first term here is the empirical distribution function of areas for the observed tree t_0, and the second term is the expected distribution function under the probability model with parameter b, our estimate of β. The Kolmogorov distance D is then the maximum vertical distance between these distribution functions. To test for significant lack-of-fit, we may use bootstrap generated trees to estimate the distribution of this test statistic under a Gibbsian null model.

APPLICATION

ILLUSTRATION OF STATISTICAL METHODOLOGY

We now look briefly at the results of one application of this model. First, in order to test the algorithm for small n, the number of grid points in a basin, we extracted 50 subnetworks using a set of 30-meter digital elevation data from Willow Creek in Montana. The networks are located in different parts of the drainage basin and are defined using data with different grid spacing, ranging from 1500 meters down to 30 meters. The number n ranges from 28 to 142, and drainage area associated with these networks ranges from 153 square kilometers down to less than 1 square kilometer. Basin elevation ranges from 1368 to 2379 meters. More specific information on these subnetworks and on the case study discussed here is given elsewhere.[20,21]

The parameter estimation algorithm failed to converge satisfactorily for 16 of these 50 networks. Shown in Table 1, for the remaining 34 networks, are the

TABLE 1 Parameter estimates b_0 for β_0 and b_1 for β_1, and 95% confidence intervals obtained by bootstrapping

Network	b_0	L_0	U_0	b_1	L_1	U_1
1	0.80	0.30	2.17	0.30	0.15	0.46
2	1.19	0.73	2.65	0.82	0.75	0.84
3	1.29	0.99	2.40	0.82	0.78	0.86
4	0.83	0.56	1.16	0.84	0.84	0.89
5	1.28	1.16	1.98	0.29	0.25	0.34
7	1.22	0.54	2.77	0.56	0.42	0.70
8	0.58	0.29	1.10	0.62	0.54	0.73
9	0.71	0.26	1.88	0.98	0.88	1.14
10	4.48	2.28	8.42	0.76	0.67	0.84
11	0.05	0.01	0.19	0.84	0.69	1.00
12	1.01	0.58	1.64	0.27	0.20	0.34
14	1.76	0.80	3.82	0.91	0.86	1.02
15	1.14	0.89	2.33	0.91	0.79	0.92
16	0.61	0.37	1.11	0.52	0.36	0.56
17	0.54	0.17	1.12	0.85	0.80	1.01
18	1.23	0.60	2.20	0.80	0.74	0.86
21	0.86	0.32	2.37	0.63	0.57	0.73
22	0.89	0.32	1.49	0.74	0.71	0.89
23	1.87	1.02	3.72	0.83	0.78	0.99
24	0.69	0.35	1.40	0.77	0.72	0.84
25	0.23	0.07	0.75	0.85	0.68	1.02
26	1.63	0.87	3.43	0.67	0.60	0.74
28	1.12	0.66	1.87	0.74	0.67	0.77
29	0.90	0.45	1.71	0.61	0.55	0.67
31	0.63	0.62	1.22	0.87	0.80	1.03
32	1.08	0.58	2.03	0.92	0.90	1.01
34	1.56	0.87	2.53	0.99	0.93	1.04
35	1.80	1.21	1.94	0.81	0.78	0.81
36	0.06	0.02	0.08	1.01	0.89	1.16
37	1.02	0.76	1.58	0.76	0.71	0.78
39	0.54	0.31	0.97	0.42	0.37	0.45
45	1.15	0.81	1.48	0.97	0.95	1.09
49	0.20	0.11	0.45	0.99	0.89	1.00
50	2.60	1.23	4.32	0.86	0.82	0.92

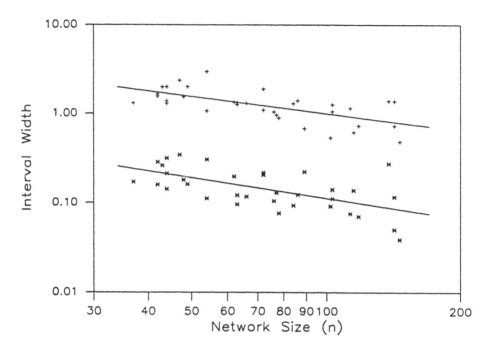

FIGURE 1 Plots of confidence interval width versus network size (n) for $\log(b_0)$ (pluses) and b_1 (asterisks).

parameter estimates for β and 95% confidence intervals (L, U) obtained using the bootstrap technique. Estimates of β_0 lie between 0.05 and 4.48, with a (geometric) average of 0.83, and estimates of β_1 are between 0.27 and 1.01 with an average of 0.75. From these confidence intervals we can test various hypotheses in which we are interested. For example, for 33 networks, estimates of β_1 are less than 1.0, and for 24 networks the 95% confidence interval excludes 1.0. Thus, for most networks by allowing β_1 to vary, we obtain a statistically significant improvement in fit over the one-parameter model in which β_1 is held at 1.0. Likewise, we see that estimates for β_1 are greater than 0.50 for 30 networks, and the 95% confidence intervals exclude 0.50 in 28 networks. Thus, evidence against the hypothesis that $\beta_1 = 0.5$ is strong. The width of the 95% confidence intervals shows a definite tendency to decrease in a power law with n. The upper plot in Figure 1 shows interval widths for $\log(\beta_0)$ and the lower plot for β_1.

In Figure 2 are shown, for a typical network (number 4), the actual network, a uniformly generated tree, trees generated from both a one- and two-parameter Gibbsian model with estimated parameter values, and a tree (an "optimal channel network") obtained by minimizing $H(s, 0.5)$. The channels in the uniform tree

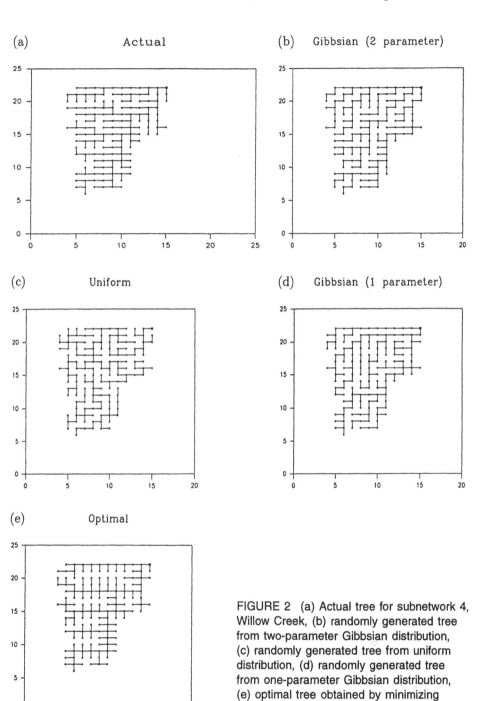

FIGURE 2   (a) Actual tree for subnetwork 4, Willow Creek, (b) randomly generated tree from two-parameter Gibbsian distribution, (c) randomly generated tree from uniform distribution, (d) randomly generated tree from one-parameter Gibbsian distribution, (e) optimal tree obtained by minimizing $H(s, 0.5)$. The outlet is the node marked with a square in the upper right corner.

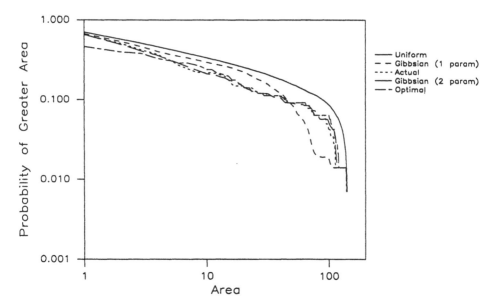

FIGURE 3 Complementary area distribution for the actual tree (subnetwork number 4) and optimal network, and mean distribution for 1000 trees generated from a uniform distribution and the one- and two-parameter Gibbsian model.

clearly tend to be too sinuous. Corresponding plots of the area distribution are shown in Figure 3. The abscissa "area" in this figure stands for number of upstream points draining into a particular grid point, and, for a given value of area, the ordinate "probability of greater area" represents the fraction of the n grid points with an associated upstream area that is greater than the given value. (Scaling properties of this distribution are discussed, for example, by Rodriguez-Iturbe et al.[13]) The distribution functions for the uniform and Gibbsian models represent averages for 1000 randomly generated trees. The probability of a greater area for the optimal channel network tends to be too low for small areas, and there is also a consistent deviation depending on magnitude of area when the one-parameter Gibbsian model is compared to the actual.

In Table 2 are shown Kolmogorov statistics D and the corresponding probability of a greater value for both one- and two-parameter Gibbsian models. These statistics provide another way of comparing the one-parameter and two-parameter models. As expected, D for the one-parameter model tends to be larger than D for the two-parameter model; the difference between these two numbers indicates the extent to which the two-parameter model fits the data better. In looking for lack of model fit, it should be noted that a test based on D would indicate significant lack of

TABLE 2 Goodness-of-fit statistic D and corresponding probability of a greater value, for both one- and two-parameter Gibbsian models

	One-parameter model		Two-parameter model	
	D	$\hat{P}$	D	$\hat{P}$
1	0.07	0.50	0.05	0.91
2	0.08	0.03	0.03	0.57
3	0.13	0.00	0.05	0.13
4	0.11	0.00	0.05	0.00
5	0.11	0.00	0.04	0.36
7	0.09	0.14	0.04	0.75
8	0.09	0.10	0.05	0.46
9	0.09	0.15	0.08	0.20
10	0.17	0.00	0.06	0.26
11	0.05	0.83	0.05	0.68
12	0.09	0.03	0.03	0.78
14	0.10	0.02	0.05	0.41
15	0.07	0.24	0.10	0.00
16	0.13	0.00	0.08	0.02
17	0.10	0.00	0.05	0.32
18	0.10	0.01	0.02	0.94
21	0.14	0.00	0.07	0.11
22	0.14	0.00	0.09	0.02
23	0.16	0.00	0.05	0.35
24	0.11	0.00	0.09	0.00
25	0.05	0.78	0.05	0.65
26	0.20	0.00	0.08	0.03
28	0.14	0.00	0.05	0.07
29	0.13	0.00	0.05	0.14
31	0.12	0.00	0.11	0.01
32	0.10	0.00	0.04	0.14
34	0.10	0.04	0.07	0.04
35	0.13	0.00	0.08	0.00
36	0.03	0.75	0.05	0.49
37	0.15	0.00	0.10	0.00
39	0.12	0.01	0.05	0.18
45	0.05	0.30	0.11	0.00
49	0.14	0.00	0.15	0.00
50	0.12	0.00	0.07	0.01

fit if the probability of a greater value is below 0.05, assuming we are testing at the 5% level. We see significant lack of fit in 25 networks for the one-parameter model, indicating rejection of this model, but in only 13 networks for the two-parameter model.

SCALING QUESTIONS

In order to look more closely at some of the scaling questions raised at the beginning of the paper, we performed another detailed study in which the Gibbsian model was again applied to networks extracted with elevation data sampled at varying grid size. Each network corresponded to one of two basins: one draining into the South Fork of Lower Willow Creek (area 99 square kilometers), and a larger one containing the first and draining into both the North Fork and the South Fork (area 204 square kilometers). Grid size ranges from 750 m down to 300 m. In Table 3 are shown estimates of the parameters β_0 and β_1 and of the scaling exponent γ in Eq. (11), computed by performing a nonlinear regression of slope against drainage area using the sampled elevation data. The estimate b_0 of β_0 shows a strong tendency to decrease as n increases. In Figure 4 is shown a scatter plot of the pairs $(b_1, -g)$, where b_1 is the estimate of the Gibbs parameter β_1 and g is the estimate of the exponent γ. A linear least-squares fit yields the equation

$$\hat{b}_1 = 1.21 + 1.19g. \tag{23}$$

We can see, however, from Figure 4 that there is considerable uncertainty in this line, especially given the small sample size, given the apparent differences between the two basins and given the great influence of the single high leverage point in the lower right. Further investigation is needed, but we feel confident of the existence of a strong connection between β_1 and γ. We note also that both β_1 and γ seem also to be related to n, so there is an elevation sampling effect in estimating both of these quantities. All we are concerned about showing here, however, is that β_1 does appear to exhibit a close connection to scaling properties of the (discretized) elevation field used to extract the network used to estimate β_1.

The correlation between estimators of β_1 and γ is fully expected in accordance with the discussion on scaling laws at the beginning of the paper, but again it is not obvious how one would proceed to check what the numerical values of these exponents tell us about discharge-area scaling. In order to investigate this question more deeply, what is needed is detailed spatial discharge data.

Finally, we emphasize once again that the Gibbsian probability model with parameters β_0 and β_1 is being used to model the two-dimensional planimetric projection of the channel network. Elevation data were used only indirectly in the estimation of β_1; i.e., the elevations were used to obtain the two-dimensional tree which in turn was used in the estimation procedure. The strong relationship between

TABLE 3 Parameter estimates b_0 for β_0 and b_1 for β_1, and estimate g for slope-area scaling exponent γ. Basins draining into both the North and South Forks, Lower Willow Creek are indicated.

Grid size	n	b_0	b_1	g
750	170	1.08	0.65	-0.51
540	352	0.79	0.67	-0.48
510	392	0.79	0.68	-0.46
480	434	0.86	0.70	-0.41
450	464	0.83	0.71	-0.45
420	513	0.69	0.67	-0.41
600[1]	623	0.77	0.48	-0.55
390	648	0.68	0.72	-0.39
360[1]	1567	0.56	0.69	-0.42
330[1]	1877	0.43	0.73	-0.40
300[1]	2265	0.33	0.77	-0.39

[1] North and South Fork, Lower Willow Creek.

γ, which is calculated directly from the elevations, and β_1, demonstrates the close tie between two- and three-dimensional drainage basin geometry. In essence, we can read scaling properties of the three-dimensional geometry from the two-dimensional network patterns.

CONCLUSIONS

We have presented methodologies for estimating the parameters of a Gibbsian probability model using the two-dimensional drainage tree obtained from digital elevation data. One of the parameters of this model, β_1, is a scaling parameter which appears, via rough hydraulic arguments, to be related to exponents in discharge/slope/area scaling relationships. In a case study using data from Willow Creek in Montana, we have shown that there is indeed a close connection between β_1 and γ, the exponent in the power law relationship between slope and drainage area. Establishing the connection between β_1 and relationships involving discharge is more problematic because the necessary discharge data is more difficult to

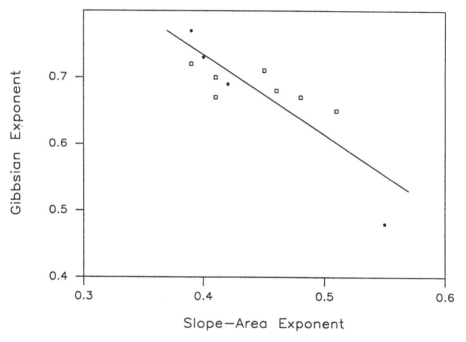

FIGURE 4 Scatter plot showing estimated slope-area exponent and Gibbsian exponent pairs $(-g, b_1)$. The open squares correspond to the South Fork, Lower Willow Creek, and the solid circles to the North and South Fork together. Also shown is the least-squares line.

obtain. It is nevertheless expected that further investigation will establish more firmly the connection between discharge scaling and network geometry as modeled by the Gibbsian distribution. Knowledge of this connection will be important in the scientific understanding of flood hydrology, and in application to problems in hazards assessment, such as the classic problem of flood prediction in ungaged basins.

REFERENCES

1. Abrahams, A. D. "Channel Networks: A Geomorphological Perspective." *Water Resour. Res.* **20** (1984): 161–188.
2. Aldous, D. "On the Markov Chain Simulation Method for Uniform Combinatorial Distributions and Simulated Annealing." *Probab. Eng. Info. Sci.* **1** (1987): 33–46.
3. Flint, J. J. "Stream Gradient as a Function of Order, Magnitude, and Discharge." *Water Resour. Res.* **10** (1974): 969–973.
4. Gupta, V. K., O. J. Mesa, and D. R. Dawdy. "Multiscaling Theory of Flood Peaks: Regional Quantile Analysis." *Water Resour. Res.* **30** (1994): 3405–3421.
5. Gupta, V. K., and E. C. Waymire. "Statistical Self-Similarity in River Networks Parametrized by Elevation." *Water Resour. Res.* **25** (1989): 463–476.
6. Karlinger, M. R., and B. M. Troutman. "A Random Spatial Network Model Based on Elementary Postulates." *Water Resour. Res.* **25** (1989): 793–798.
7. Leopold, L. B., and T. Maddock, Jr. "The Hydraulic Geometry of Stream Channels and Some Physiographic Implications." Prof. Paper 252, U.S. Geological Survey, 1953.
8. Leopold, L. B., and M. G. Wolman. "River Channel Patterns: Braided, Meandering, and Straight." Prof. Paper 282-B, U.S. Geological Survey, 1957.
9. Leopold, L. B., M. G. Wolman, and J. P. Miller. *Fluvial Process in Geomorphology.* New York: W. H. Freeman, 1964.
10. Lichty, R. W., and M. R. Karlinger. "Climate Factor for Small-Basin Flood Frequency." *Water Res. Bull.* **26** (1990): 577–586.
11. Rinaldo, A., I. Rodriguez-Iturbe, R. Rigon, R. L Bras, E. Ijjasz-Vasquez, and A. Marani. "Minimum Energy and Fractal Structures of Drainage Networks." *Water Resour. Res.* **28** (1992): 2183–2195.
12. Rinaldo, A., I. Rodriguez-Iturbe, R. Rigon, E. Ijjasz-Vasquez, and R. L. Bras. "Self-Organized Fractal River Networks." *Phys. Rev. Lett.* **70** (1993): 822–825.
13. Rodriguez-Iturbe, I., E. Ijjasz-Vasquez, R. L. Bras, and D. G. Tarboton. "Power Law Distributions of Discharge Mass and Energy in River Basins." *Water Resour. Res.* **28** (1992a): 1089–1094.
14. Rodriguez-Iturbe, I., A. Rinaldo, R. Rigon, R. L. Bras, A. Marani, and E. Ijjasz-Vasquez. "Energy Dissipation, Runoff Production, and the Three-Dimensional Structure of River Basins." *Water Resour. Res.* **28** (1992b): 1095–1103.
15. Shreve, R. L. "Infinite Topologically Random Channel Networks." *J. Geol.* **75** (1967): 179–186.
16. Smith, J. A. "Representation of Basin Scale in Flood Peak Distributions." *Water Resour. Res.* **28** (1992): 2993–2999.

17. Sun, T., P. Meakin, and T. Jossang. "The Topography of Optimal Drainage Basins." *Water Resour. Res.* **30** (1994): 2599–2610.

18. Tarboton, D. G., R. L. Bras, and I. Rodriguez-Iturbe. "Scaling and Elevation in River Networks." *Water Resour. Res.* **25** (1989): 2037–2051.

19. Tasker, G. D., and J. R. Stedinger. "An Operational GLS Model for Hydrologic Regression." *J. Hydrol.* **111** (1989): 361–375.

20. Troutman, B. M., and M. R. Karlinger. "Gibbs' Distribution on Drainage Networks." *Water Resour. Res.* **28** (1992): 563–577.

21. Troutman, B. M., and M. R. Karlinger. "Inference for a Generalized Gibbsian Distribution on Channel Networks." *Water Resour. Res.* **30** (1994): 2325–2338.

22. Wolman, M. G. "The Natural Channel of Brandywine Creek, Pennsylvania." Prof. Paper 271, U.S. Geological Survey, 1955.

Daniel H. Rothman and John P. Grotzinger
Department of Earth, Atmospheric, and Planetary Sciences, Massachusetts Institute of Technology, Cambridge, MA 02139

Thickness Statistics of Sedimentary Layers Generated by Gravity-Driven Flows

This paper originally appeared as "Scaling Properties of Gravity-Driven Sediments." In *Nonlinear Processes in Geophysics*, Vol. 2, 178–185. Copyright © 1995 by Publisher; reprinted, with minor changes, by permission.

Recent field observations of the statistical distribution of turbidite and debris flow deposits are discussed. In some cases one finds a good fit over 1.5–2 orders of magnitude to the scaling law $N(h) \propto h^{-B}$, where $N(h)$ is the number of layers thicker than h. Observations show that the scaling exponent B varies widely from deposit to deposit, ranging from about $1/2$ to 2. Moreover, one case is characterized by a sharp crossover in which B increases by a factor of two as h increases past a critical thickness. We propose that the variations in B, either regional or within the same deposit, are indicative of the geometry of the sedimentary basin and the rheological properties of the original gravity-driven flow. The origin of the power-law distribution remains an open question.

INTRODUCTION

Gravity-driven sedimentation in oceanic basins occurs as the result of slumping, or avalanche, events at the edge of the continental shelf.[17,19] These slumping events, which are related in part to landslides and tsunamis, originate on relatively steeply sloped submarine topography. They create subaqueous flows known as gravity currents[25] that can flow for hundreds of kilometers or more. Once these flows finally lose energy and stop, they deposit the sediment that is no longer mobilized by the flow. The historical record of these gravity-driven sedimentation events is the sedimentary succession itself.

Gravity-driven sedimentation has been the subject of much study during the last half century,[13,15,16,18] in part because its discovery helped explain the distribution of sediment in oceanic basins. One outcome of this work has been a classification of sediment types based roughly on the role that turbulence plays in the transport of the sediment.[17] At the turbulent end of the spectrum are *turbidity currents*, the deposits of which are called *turbidites*. At the laminar end of the spectrum are *debris flows*. Whereas in the former case the transport of sediment is supported by the upward motion of turbulent, turbid vortices, the latter type of sediment support is thought to be due to the intrinsic strength of a relatively thick mixture of sand and debris.

Recent studies have revealed some interesting statistical properties of turbidites.[10,11,22,23] For each of the turbidite successions studied in these papers, if one measures the thickness h of all the layers, then the number of layers thicker than h scales like h^{-B} above a small thickness cutoff, with $B \approx 1$. Such a scaling property is significant for several reasons.

First, the appearance of a power law indicates that the dynamical mechanism responsible for turbidite deposition may be scale invariant. This in turn is suggestive[22] of the many recent studies on "self-organized criticality,"[1] which is itself an attempt to formulate a theory for the ubiquitous occurrence of scale invariance in nature. The connection with self-organized criticality is due in large part to the fact that Bak et al. used avalanches as the archetypal example of their theory.

Second, one would like to know whether power-law scaling for turbidites represents an intrinsic dynamical property of turbidite systems themselves, or whether it is just the signature of power-law scaling for an external, causal mechanism. The initiation of turbidity currents is related to slope instabilities, and could be influenced by floods[27] and earthquakes,[2,11] each of which exhibit power-law size distributions. However, the power-law scaling in all these systems suggests that there may be a generic physics describing them all. Such a generalization is indeed the objective of the proponents of self-organized criticality.

Last but not least is the importance of such a scaling law for geology. Specifically, the power law may be viewed merely as a statistical distribution with a parameter B. One is then led to the following questions:

1. Is the power law generic?
2. If so, is there a typical or generic value for B?
3. If there is no generic value for B, what geologic factors determine its value?

It is the objective of this chapter to provide preliminary answers to these questions. To achieve such a goal, we look at debris flows in addition to turbidite deposits. Our results indicate that the power law is not generic; moreover, there is no generic value for B in the cases in which power laws exist. Like most other real-world phenomena, the problem is plainly more complex than idealized theories would lead one to presume. Nevertheless, we believe that there remains some order to extract from this complexity. Specifically, for the cases in which power laws exist, we propose that the value of B is determined by factors related to both the geometry of sediment deposition and the rheology of the gravity-driven flow. By doing so, we leave open the possibility that generic mechanisms govern the dynamics; the available data, however, can neither support nor deny such a hypothesis. Moreover, the factors that determine whether one finds a power law remain a matter of speculation.

In what follows we first provide a brief qualitative introduction to gravity-driven sedimentation. We then discuss the statistical distributions of turbidite and debris flow sediments in three distinct geological settings. Lastly, we attempt to "rationalize" our empirical findings by proposing elementary scaling laws, for which aspects of basin geometry and flow rheology play a role via the introduction of a parameter.

GRAVITY-DRIVEN SEDIMENTATION: A CARTOON VIEW

In this brief section we give an overview of gravity-driven sedimentation for the nonspecialist. Our intention is only to provide the geologic context of this subject. Gravity currents themselves, however, are the object of much study; they are not only ubiquitous in nature (in both atmospheric and oceanographic flows), but are also of significant engineering interest, for which oil spills are but one example. A beautiful introduction to the theory and phenomenology of the general subject, along with numerous references and photographs, is given in the recent book by Simpson.[25] More detailed recent theoretical and experimental studies can be found in the papers by Huppert and his colleagues at Cambridge[4,5,12] some results of which we will refer to later.

Figure 1 gives the general setting. Due to transport by traction or in suspension, sediment is deposited at the edge of the continental shelf and along the continental slope, with a volume flux Q. Because the slope is unstable, avalanche-like events known as *slumps* occur at widely spaced, discrete intervals of time. The slump creates a region of dense fluid (i.e., a mixture of water and sediment), which is gravitationally unstable with respect to the "pure," less dense, fluid below it.

Thus the slump develops into a *gravity current* which flows along the sea bottom. Eventually the current slows down when it reaches the relatively flat basin plain, at which point the sediment it has carried finally settles out.

The slumping events are rapid—from beginning to end, they can last from about one minute to one day, depending on the size and the range of the flow— but they occur rarely—the time interval between events can range from years to thousands of years. This low frequency of events helps in distinguishing individual

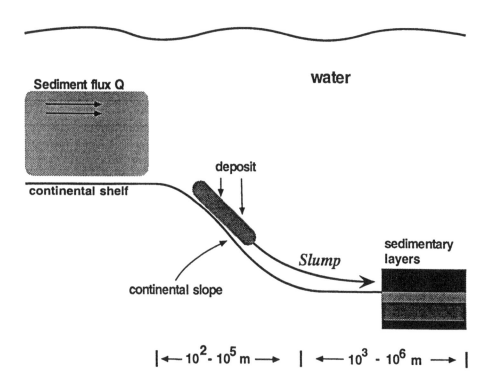

FIGURE 1   Typical setting for gravity-driven sedimentation in oceanic basins. A flux Q of sediment is deposited by traction or in suspension on the continental slope. At widely spaced time intervals, this sediment slumps in an avalanche-like event, creating a gravity current that flows out to great distances. The sediment that had been mobilized by the gravity current settles out in the basin plain, creating (in ideal circumstances) a well-defined layer. Depending on the type of current, the layer could be characterized as either a turbidite deposit or a debris flow. Often, thin mud layers lie between layers as evidence of a constant "background" sedimentation unassociated with gravity-driven deposits.

layers in field situations, because unrelated, mud-sized sediment falls constantly and independently in the "background," creating a thin and distinctive mud layer between each slumping event.

Middleton and Hampton[17] have classified subaqueous sediment gravity flows according to the composition of the sediment, the rheology of the flow, and the degree to which the flow is turbulent. Their classification is purely qualitative and left open to interpretation by individual field geologists, but it is useful nonetheless. To simplify matters, we consider only the two extreme cases.

At the turbulent end of the spectrum are turbidity currents. Here the density of the flow is not much greater than that of the sea water, and the sediment remains mobilized due to the action of powerful vortices at the head of the flow. The deposits are characterized by a graded sequence in which the largest clasts (i.e., pebble-size sediment) fall to the bottom whereas the finest-grade sediment settles last, and is thus found on top. A schematic view of such an ideal turbidite layer is shown in Figure 2(a).

TURBIDITE
(a)

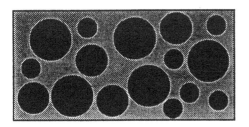

DEBRIS FLOW
(b)

FIGURE 2   (a) Distribution of grain size in a typical turbidite deposit. (b) Distribution of grain size in a typical debris flow deposit. In the turbidite the grains are in contact with each other and no supporting matrix is present. In contrast, in the debris flow deposit the grains are supported by interstitial mud matrix, here colored gray.

At the laminar end of the spectrum are debris flows. In this case, the flowing mixture of sediment and water can be an extremely dense mud with a non-Newtonian rheology. The salient property of the mud is that it has a finite yield strength; i.e., it flows only if subjected to a sufficiently large shear stress. Middleton and Hampton say that "debris flow essentially resembles flow of wet concrete."[17] Both the density of the mud and its finite yield strength give debris flows distinct characteristics that may be identified by a field geologist. Specifically, and in contrast to turbidites, graded sequences of clasts are infrequent, and the clasts are distributed widely in space. Indeed, large boulders can appear to have been levitated by the flow, because the carrying fluid (1) is nearly as dense as the boulder itself, and (2) acts like a solid if insufficiently stressed. A schematic view of an ideal debris flow layer is shown in Figure 2(b).

FIELD OBSERVATIONS

Below we describe three gravity-driven deposits that we have studied by direct observation in the field. The first, from southeast California, allows us to directly compare the statistical distribution of turbidites and debris flows in the same deposit. The second, from turbidites in Karoo Basin, South Africa, displays an intriguing crossover from one power law to another. The third, from Barberton, South Africa, does not conform well to a power-law distribution.

KINGSTON PEAK FORMATION, SOUTHEAST CALIFORNIA

Turbidites and debris flows of the Kingston Peak Formation in southeast California are approximately 700 million years old and accumulated in a narrow fault-bounded trough.[22] Turbidites are well expressed as even beds up to 2 m thick which show minimal evidence of erosion along their bases. Also, amalgamation of individual turbidites is rare. Debris flows are much thicker (up to 10 m) and commonly show matrix support of clasts which range up to 0.48 m in diameter. In some cases debris flows are amalgamated.

In a previous study, we examined only turbidites from this area.[22] We found a reasonably good fit to the power law $N(h) \propto h^{-B}$, where $N(h)$ is the number of layers thicker than h, and $B = 1.39 \pm 0.02$. The total number of turbidite layers was 1235, and the fit was over approximately two orders of magnitude in h, ranging from centimeters to meters.

Interspersed among those turbidite layers are also 24 debris flows. These debris flows were qualitatively distinguished from turbidites according to the schematic diagrams of Figure 2. Figure 3 is a log-log plot of $N(h)$ for the 1235 turbidites compared to $N(h)$ for the 24 debris flows. Three qualitative differences between the two plots are evident. First, there are more than 50 times as many turbidites as debris flows. Second, debris flows range in thickness from 10^1 to 10^3 cm, whereas turbidites are scattered from 10^0 to 10^2 cm. The third difference is equally unsubtle, but is the most interesting: the debris flows scale with $B = 0.49 \pm 0.01$, which is roughly *one-third* of the B-value found for the turbidites. Why such a dramatic difference in B-values could exist within the same sedimentary succession is considered further below.

LAINGSBURG FORMATION, KAROO BASIN, SOUTH AFRICA

Turbidites of the upper Laingsburg Formation (Karoo Basin) were measured along the Buffels River, north of the town of Laingsburg, South Africa. The Laingsburg Formation is about 275 million years old and its turbidites have been described by Bouma.[3] Most turbidites are even bedded and only the thicker beds show frequent amalgamation. Most turbidites are laterally continuous for distances greater than 100 m. Deposition occurred within the confines of a tectonically active foreland basin developed in front of an advancing mountain belt.

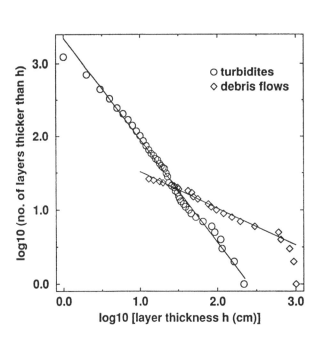

FIGURE 3 Logarithm (base 10) of the number of layers thicker than h as a function of the logarithm of layer thickness h, for 1235 turbidites (circles) and 24 debris flows (diamonds) observed in the Kingston Peak Formation, SE California. The turbidite data are compared to a straight line with slope $-B = -1.39$, the best fit to a linear regression computed from all points except the ones for the smallest and largest values of h. The debris flow data are compared to a straight line with slope $-B = -0.49$, the best fit to a linear regression computed from all points except those corresponding to the three thickest layers.

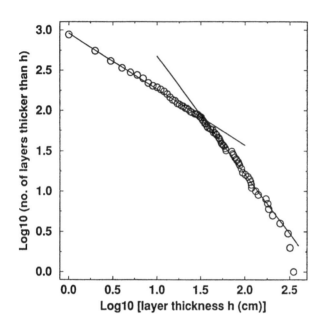

FIGURE 4 Logarithm (base 10) of the number of layers thicker than h as a function of the logarithm of layer thickness h, for 878 turbidites from Laingsburg Formation, Karoo, South Africa. The thin beds fall roughly on a line with slope $-B = -0.70$, the best fit to a linear regression computed from all points corresponding to $h < 10^{1.5} \approx$ 30 cm. The thick beds fall roughly on a line with slope $-B = -1.47$, the best fit to a linear regression computed from all points corresponding to $h > 10^{1.5}$ except for the two largest values of h.

Figure 4 is a plot of $N(h)$ for 878 turbidites from this formation. One finds that, for $10^0 \leq h \leq 10^{1.5} \approx 30$ cm, there is a good fit to the power-law scaling with $B = 0.70 \pm 0.01$. However, for $10^{1.5} \leq h \leq 10^{2.5}$ cm, the fit is for a value of B more than twice as large; specifically, $B = 1.47 \pm 0.02$. Further investigation of the data shows that the thick layers are apparently randomly situated among the thin layers. For example, if the data set is divided into two halves, in which the first half has the first 439 layers and the second half the rest, then $N(h)$ for each set still looks qualitatively similar to Figure 4. Thus, unlike the turbidites of southeast California, we find two distinct scaling regimes rather than just one.

FIG TREE GROUP, BARBERTON, SOUTH AFRICA

Turbidites of the Fig Tree Group were measured along a road cut south of the Sheba Mine within the Barberton Mountain Land, South Africa. The Fig Tree Group is approximately 3.2 billion years old and contains some of the oldest turbidites on earth. These have been well described by Eriksson,[8] who interpreted deposition to have taken place on the middle part of a submarine fan, both within and between channels. Amalgamation of turbidites is common at the study site. Deposition probably occurred in a tectonically active basin, although the exact type is uncertain.[7,9,20]

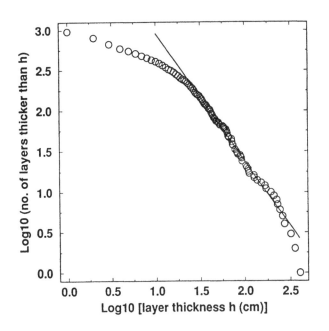

FIGURE 5   Logarithm (base 10) of the number of layers thicker than h as a function of the logarithm of layer thickness h, for 962 turbidites from Fig Tree Group, Barberton, South Africa. The data do not conform to either a simple power law or a clean crossover as in Figure 4. One finds, however, approximate power-law scaling for the thicker turbidite layers; the straight line has slope $-B = -1.58$, which is the best fit to a linear regression computed from the points corresponding to $h > 10^{1.4} \approx 25$ cm, except for the two largest values of h.

Figure 5 is a plot of $N(h)$ for 962 turbidites from this section. There is generally a poor fit to a power law, although the layers thicker than about 25 cm appear to scale with $B = 1.58 \pm 0.02$ over about one order of magnitude. There are qualitative similarities to the Karoo data, but there is no well-defined crossover as was found in that case.

DISCUSSION

From the three datasets of Figures 3, 4, and 5, two conclusions may be drawn. First, simple one-parameter scaling behavior is not always observed. Second, if there is power-law scaling, the exponent B appears to depend on flow type—that is, whether the deposit is a turbidite or debris flow—or, in the case of Figure 4, the magnitude of the layer thickness. Whereas the origin of the power-law scaling is difficult to pin down, the dependence of the exponent on the type or size of the flow appears considerably easier to address. Below we consider some possible scaling laws. Our analysis is similar but simpler and more physical than our previous work.[22,23] A related study has also been recently reported by Malinverno.[14]

SCALING LAWS

Because our datasets are small, we have displayed only the (smoother) cumulative distributions $N(h)$. They may be related to a frequency distribution $\rho_h(h)$ by

$$\rho_h(h) = -\frac{1}{\Delta t}\frac{d}{dh}N(h).\tag{1}$$

Here $\rho_h(h)$ is the number of layers of thickness h deposited per unit time and Δt is the duration of geologic time from the bottom to the top of the section. For the cases in which $N(h) \propto h^{-B}$, we have

$$\rho_h(h) \propto h^{-B-1}.\tag{2}$$

Although we can measure only the thickness distribution of sedimentation events, we would like to know the *volume* distribution so that we may distinguish aspects of basin and flow geometry from intrinsic dynamical processes on the slope. We assume that the thickness h is approximately uniform throughout a deposit covering an area S; thus the volume $V = Sh$. Additionally, we assume that h scales with V according to

$$h \propto V^\alpha.\tag{3}$$

We call α the *spreading exponent* and expect $0 \leq \alpha \leq 1$. Consideration of some special cases gives some insight into the spreading:

1. $\alpha = 0$. *Perfect spreading.* $S \propto V$ and all layers have the same thickness.
2. $\alpha = 1/3$. *Self-similar areal spreading.* If spreading is in two dimensions, then $S \propto V^{2/3}$, and all three linear dimensions (length, width, and height) respond roughly equally to changes in V.
3. $\alpha = 1/2$. *Self-similar channelized spreading.* If spreading is confined to a channel, and thus only one dimension, then both the height and the length depend equally on V.
4. $\alpha = 1$. *No spreading.* Each sedimentation event spreads over the same area S.

Interestingly, an empirical study of stursztoms (rock slides) appears to have found self-similar areal spreading with $\alpha \approx 1/3$.[6] There is no reason to expect self-similar spreading in either one or two dimensions, however. In a study of channelized turbidity flows, Dade and Huppert[5] predict $\alpha = 2/5$. Here, however, we are considering deposits ranging from viscous non-Newtonian debris flows to turbulent turbidity currents in unknown geometries. In this continuum of rheologies and geometries, we simply expect that as the flow becomes less like a debris flow and less confined, the tendency to spread increases, and thus the value of α decreases. A detailed analysis for non-Newtonian rheologies, à la Dade and Huppert, would nevertheless be necessary to prove this point and make it precise.

Given Eq. (3), a change of variables from h to V in Eq. (4) gives the number of events of volume V per unit time:

$$\rho_V(V) = \rho_h[h(V)]\frac{dh}{dV}$$
$$= AV^{-\alpha B - 1}. \tag{4}$$

The prefactor A is related to the volume flux Q of sediment to the continental slope (see Figure 1). Specifically, we define Q to be the volume of sediment delivered to the shelf per unit time. Clearly, A depends on Q, and, naively, one would expect $A \propto Q$. The dependence of A on the size of the system should also be considered, however. For example, although sediment may be delivered uniformly to the continental slope at rate Q, only a subset of the slope may be "active" in the sense that it would channel sediment into a particular part of the basin plain. Typically these active regions would correspond to submarine canyons, and they may not fill the shelf-break uniformly. More specifically, as in analogous continental drainage networks,[21,26] the active drainage areas on the continental slope may form a fractal set. (Indeed, analysis of high-resolution digital bathymetric maps of submarine canyon networks supports such a conclusion.[24]) In this case one would expect the more general relation

$$A \propto QV_{\max}^{-\nu}, \tag{5}$$

where $V_{\max}$ is the largest possible event (i.e., the system size) and ν is related to the fractal dimension of the active region. In particular, on a two-dimensional map, the fractal dimension of the active region would be $2 - \nu$, with $0 \leq \nu \leq 1$. We return to a discussion of ν below.

An assumption of a statistically stationary state—specifically, that Q is roughly constant in a coarse-grained sense over the time period Δt—allows one to relate the flux to the volume frequency distribution:

$$Q = \int_{V_{\min}}^{V_{\max}} V\rho_V(V)dV. \tag{6}$$

Here $V_{\min}$ is the smallest possible sedimentation event. Substitution of Eq. (4) gives

$$Q \propto A \cdot \left(V_{\max}^{1-\alpha B} - V_{\min}^{1-\alpha B}\right). \tag{6}$$

By letting $V_{\min} \to 0$ and substituting Eq. (5) for A we can eliminate Q to obtain

$$V_{\max}^{1-\nu-\alpha B} = \text{constant}, \qquad \alpha B < 1. \tag{7}$$

The inequality $\alpha B < 1$ is required for Q to be finite. For Eq. (7) to hold, the dependence of the left-hand side on $V_{\max}$ must vanish. Thus we find that α, ν, and B are related by

$$B = \frac{1-\nu}{\alpha}. \tag{8}$$

Note that the rigorous bound $\alpha B < 1$ requires $\nu > 0$. If the power law were not applicable in the limit $V_{\min} \to 0$, the bound $\alpha B < 1$ would not be required but Eq. (8) would still be valid if $V_{\max}^{1-\alpha B} \gg V_{\min}^{1-\alpha B}$.

INTERPRETATION

Since B is measured, then if ν is known, Eq. (8) may be used to infer the value of the spreading exponent α—thus providing quantitative insight into qualitative characteristics of the original flow. However, as we show below, knowledge of ν is not necessary to make purely qualitative conclusions.

In Table 1 we give the results that would follow for the turbidites and debris flows of Figure 3 and the crossover behavior in Figure 4. Since ν is not known, we express α in terms of the *reduced spreading exponent* $\alpha/(1-\nu) = B^{-1}$. The principal conclusion, given in the rightmost column, concerns the relative spreading of the two types of flows found in each formation. The determination of the relative spreading does not depend on the value of ν, but it does require that ν be constant within the same formation.

For the case of SE California, the scaling theory predicts that the debris flows spread less (i.e., that α is larger) compared to the turbidites. For example, if $\nu = 1/2$, then one finds that $\alpha_{\text{debris}} = 1$ and $\alpha_{\text{turbidite}} = 0.4$. The former value would lead to the conclusion that the debris flows did not spread at all. The latter value would coincide with the Dade-Huppert prediction for channelized turbidity flows.[5]

The case of the Karoo turbidites (Figure 4) has a more subtle interpretation. The thin-layer part of the curve appears to represent flows that had much less of tendency to spread than the thick-layer part of the curve. The presence of two such spreading characteristics within the same formation could be explained by two different sources of turbidity flows. The source of the thin layers would be relatively closer to the measured formation than the source of the thick layers. The absence of thin layers from the far (spreading) source would be due to the thin layers not having travelled sufficiently far to reach the measured outcrop. The absence of thick layers from the near source might indicate that the near source was fed by a smaller drainage system, with a smaller maximum event size.

TABLE 1 The exponents B and reduced spreading exponent $\alpha/(1 - \nu)$ for the distributions displayed in Figures 3 and 4. The characterization of spreading as "much" or "little" is purely relative; it assumes that ν is the same for each type of flow within the same formation, and depends only on the relative values of B.

Location	Deposit type	B	$\alpha/(1 - \nu)$	Spreading
SE Calif.	debris flow	0.5	2.0	little
	turbidite	1.4	0.7	much
Karoo	thin layer	0.7	1.4	little
	thick layer	1.5	0.7	much

In closing this section, it is worthwhile to comment on the role of ν. If ν were approximately zero, as the simplest considerations would predict, then the value of α that would be inferred from the debris flows from SE California and the thin layers from Karoo would be physically implausible ($\alpha > 1$). Thus it appears that $\nu > 0$ may be a necessary component of the theory. However, the data may be too scarce to make such a conclusion or the scaling theory may be too simple to have any general validity.

CONCLUSIONS

We have two principal conclusions. First, power-law size distributions are not ubiquitous among gravity-driven sediments. Second, when power laws do exist, the scaling exponents can vary widely from formation to formation.

This chapter has primarily addressed the second conclusion. The fact that the scaling exponents differ means that the size distributions are *non-universal*; i.e., no generic size distribution exists when measured in terms of thicknesses. Here, we have exploited the differences among power-law thickness distributions to infer qualitative characteristics of the tendency for deposits to spread after slumping. However, it may still be possible that the product αB and the volume frequency distribution $\rho_V(V) \propto V^{-\alpha B - 1}$ could be universal. Indeed, in our analysis, such would be the case if the exponent ν were everywhere the same.

Perhaps the most important question, however, concerns the origin of the power-law distributions, and why they are not always observed. If external mechanisms such as earthquakes or floods were the dominant cause of sedimentation events, the likelihood or lack thereof of these external factors would explain regional variations. However, one would still need to ask why the size distribution of these external events were such that power-law sedimentation events were produced. As an indication of the complexity of all such questions, we note that if Eq. (3) did not hold—i.e., if thickness were not necessarily a power-law function of volume—then one could not observationally distinguish between power-law and non-power-law volume distributions. Thus, from the available data, it remains possible that volume distributions are generically distributed as a power law, but that geometric aspects of the depositional basin can remove any indication of it. Only further work, in the form of theory, observation, and experiment, can help resolve these issues.

ACKNOWLEDGMENTS

This work was partially supported by NSF Grants 9218819-EAR and 9058199-EAR. We thank Jen Carlson for her assistance in processing the South African turbidite data.

REFERENCES

1. Bak, P., C. Tang, and K. Wiesenfeld. "Self-Organized Criticality." *Phys. Rev. A* **38** (1992): 363–374.
2. Beattie, P., and W. B. Dade. "Is Scaling in Turbidite Deposition Consistent with Forcing by Earthquakes?" Preprint, 1994.
3. Bouma, A. H., and H. Wickens. "Permian Passive Margin Submarine Fan Complex, Karoo Basin, South Africa: Possible Model of Gulf of Mexico." *Transactions—Gulf Coast Assoc. Soc.* **41** (1991): 30–32.
4. Dade, W. B., and H. E. Huppert. "A Box Model for Non-entraining, Suspension Driven Gravity Surge on Horizontal Surfaces." *Sedimentology* (1994): in press.
5. Dade, W. B., and H. E. Huppert. "Predicting the Geometry of Deep-Sea Turbidites." *Geology* (1994): in press.
6. Davies, T. R. H. "Spreading of Rock Avalanche Debris by Mechanical Fluidization." *Rock Mechanics* **15** (1982): 9–24.
7. de Wit, M. J., C. Roering, R. J. Hart, R. A. Armstrong, C. E. J. de Ronde, R. W. E. Green, M. Tredoux, E. Peberdy, and R. A. Hart. "Formation of Archaean Continent." *Nature* **357** (1992): 553–562.
8. Eriksson, K. A. "Hydrodynamic and Paleogeographic Interpretation of Turbidite Deposits of the Archean Fig Tree Group of the Barberton Mountain Land, South Africa." *Geol. Soc. Am. Bull.* **91** (1980): 21–26.
9. Eriksson, K. A. "Transitional Sedimentation Styles in the Moodies and Fig Tree Groups, Barberton Mountain Land, South Africa: Evidence Favouring an Archean Continental Margin." *Precambrian Res.* **12** (1980): 141–160.
10. Hiscott, R., A. Colella, P. Pezard, M. Lovell, and A. Malinverno. "Sedimentology of Deep Water Volcanoclastics, Oligocene Izu-bonin Forearc Basin, Based on Formation Microscanner Images." *Proc. Ocean Drilling Program, Sci. Results* **126** (1992): 75–96.
11. Hiscott, R., and J. Firth. "Scaling in Turbidite Deposition—Discussion." *J. Sedimentary Res.* **A64** (1994): 933.
12. Huppert, H. E., and J. E. Simpson. "The Slumping of Gravity Currents." *J. Fluid Mech.* **99** (1980): 785–799.
13. Kuenen, P., and C. I. Migliorini. "Turbidity Currents as a Cause of Graded Bedding." *J. Geology* **58** (1950): 91–127.

14. Malinverno, A. "Scaling Exponents of Measured Turbidite Bed Thicknesses and Turbidite Bed Volumes." In *EOS, Transactions, American Geophysical Union 1994 Spring Meeting*, 351. American Geophysical Union, 1994.
15. Middleton, G. V. "Experiments on Density and Turbidity Currents, I. Motion of the Head." *Canad. J. Earth Sci.* **3** (1966): 523–546.
16. Middleton, G. V. "Experiments on Density and Turbidity Currents, II. Uniform Flow of Density Currents." *Canad. J. Earth Sci.* **3** (1966): 627–637.
17. Middleton, G. V., and M. A. Hampton. "Subaqueous Sediment Transport and Deposition by Sediment Gravity Flows." In *Marine Sediment Transport and Environmental Management*, edited by D. J. Stanley and D. J. P. Swift, 197–218. New York: Wiley, 1976.
18. Middleton, G. V. "Experiments on Density and Turbidity Currents, III. Deposition of Sediment." *Can J. Earth Sci.* **4** (1966): 475–505.
19. Middleton, G. V. "Sediment Deposition From Turbidity Currents." *Ann. Rev. Earth Planet. Sci.* **21** (1993): 89–114.
20. Nocita, B. W., and D. Lowe. "Fan-Delta Sequence in the Archean Fig Tree Group, Barberton Greenstone Belt, South Africa." *Precambrian Res.* **48** (1990): 375–393.
21. Rodriguez-Iturbe, I., E. J. Ijjász-Vásquez, R. Bras, and D. Tarboton. "Power-Law Distributions of Discharge Mass and Energy in River Basins." *Water Resour. Res.* **28** (1992): 1089–1093.
22. Rothman, D. H., J. P. Grotzinger, and P. B. Flemings. "Scaling in Turbidite Deposition." *J. Sedimentary Res.* **A64** (1994): 59–67.
23. Rothman, D. H., J. P. Grotzinger, and P. B. Flemings. "Scaling in Turbidite Deposition—Reply." *J. Sedimentary Res.* **A64** (1994): 934.
24. Rothman, D. H., and J. P. Grotzinger. "Dynamical and Geometrical Aspects of the Continental Shelf." Unpublished manuscript, 1994.
25. Simpson, J. E. *Gravity Currents: In the Environment and the Laboratory.* Chichester: Ellis Horwood Limited, 1987.
26. Tarboton, D., R. Bras, and I. Rodriguez-Iturbe. "The Fractal Nature of River Networks." *Water Resour. Res.* **24** (1988): 1317–1322.
27. Turcotte, D. L. "Fractal Sspects of Geomorphic and Stratigraphic Processes." *GSA Today* **4** (1994): 201–213.

Earthquakes

Stephen G. Eubank
Santa Fe Institute, 1399 Hyde Park Road, Santa Fe, NM 87501

Thoughts on Modeling and Prediction of Earthquakes

1. INTRODUCTION

It is generally agreed that no clear precursors of earthquakes or predictive models have been identified. This paper concerns three questions which arise from that observation:

- What kinds of models are we looking for?
- How do dynamics determine the types of precursors and models to look for?
- Why haven't we seen clear precursors yet?

In addressing the first question, I will describe an approach to model building applicable to complex systems. I offer the view of someone familiar with the variety of behaviors possible in a nonlinear dynamical system and the capabilities and limitations of time series models. I hope this will bring a different perspective to the field and encourage a reevaluation of current modeling techniques. As an outsider to the field I do not presume to attempt to survey the literature on earthquake modeling and prediction. For a timely review in the spirit of this article, see Gabrielov and Newman.[4]

2. WHAT KINDS OF MODELS ARE NEEDED

2.1 WHAT IS A MODEL?

The traditional approach to modeling in physics has been to build "bottom up" models which attempt to explain large-scale behavior in terms of well-known physics on small scales. The model is considered useful if it identifies general dynamics which can be specialized to a particular system. A typical model produced by this approach is a set of differential equations (capturing the general dynamics) which can be specialized by specifying boundary conditions. Some models may not require specialization. For example, scaling laws near phase transitions capture universal aspects of the large-scale behavior.

In complex systems, however, the arena in which dynamics takes place often evolves along with the system itself. In such systems the traditional approach is not always fruitful because the separation between boundary conditions and dynamics is not clear cut. For instance, suppose we were given a model which could predict seismic activity based on knowledge of the strain field and fault geometry. It seems as though this is exactly the clean separation between boundary conditions and dynamics that we want, but how clean is it? The dynamics, after all, is responsible for the geometry of the faults and the dynamics and geometry together are responsible for the strain field. To make a model for a particular fault zone requires at least three additional assumptions:

- There is a length scale below which changes in fault geometry do not affect the dynamics.
- The system is isolated—that is, there is a length scale above which distant shocks do not affect the strain field or fault geometry.
- It is possible to measure the geometry and strain field with the required precision within the time constraints imposed.

These are precisely the kinds of assumptions that break down in many complex systems. Even worse, we may find that some aspect of the dynamics itself (for example, the existence of certain interaction terms) depends on what are being called boundary conditions.

On a more optimistic note, there is evidence that complex systems will self-organize into exactly the sort of critical regimes for which traditional methods can find universal behavior.[1] The question then becomes whether these universal behaviors are useful for hazard reduction or mitigation.

This chapter focuses on an entirely different sort of model—data-driven models provided by statistical or time series analysis. The goal of these models is to predict the future behavior of the particular time series under study. They make no attempt to split dynamics apart from boundary conditions or to explain the behavior in terms of fundamental physical laws. It may be possible to extract properties such as the most important inputs or time scales from several such models, but that is not the main goal.

The advantage of using data-driven models in complex systems is that they provide a direct answer to the question: "Is there a pattern in these observations which can be exploited for prediction?" By choosing the location and time scale of the observations the modelbuilder has control over the meaning and usefulness of any patterns found. The essential assumption made in using these models for prediction is stationarity. While this may not be a valid assumption in complex systems, it is hard to imagine any predictive model which does not rely on stationarity in some form or another. Finally, there is no need to specialize the model, since it is tuned to the observations it was built on. This is, of course, a disadvantage as well, since finding a model for one time series does not necessarily guide research on others. The hope is that models for different data sets can be built using similar techniques.

An excellent example of the data-driven approach is Rob Shaw's analysis of a dripping faucet.[7] The problem is to understand the pattern of time intervals between drops in a leaky faucet. (Compare, for instance, the problem of identifying "characteristic earthquakes.") It is possible to model this system in terms of surface tension, gravity, and flow rate, but the resulting model does not give as clear a picture of the overall phenomenon as a simple model based directly on the observed time interval between drops. Other examples from the field of earthquake prediction are those by Keilis-Borok.[6]

In the following discussion, I will use the following three forecasts as typical examples of the forecasts a data-driven model might make:

1. There will be a magnitude 8.0 earthquake along (some particular section of) the San Andreas fault within the next 100 years.
2. This seismograph will register...waves of .2 g amplitude....
3. A large quake will occur in the Los Angeles basin within the next two days.

2.2 HOW USEFUL IS A MODEL?

This question assumes we have agreed upon a criterion for assessing utility. The criterion suggested by this workshop's title is that a good model should help in hazard mitigation. As noted above, this is different from understanding the causes of seismic activity as an end in itself.

The most important attributes for characterizing prediction techniques for natural hazards are precision and accuracy in space, time, and magnitude. The location of the model in this space determines the practical mitigation strategies. It is thus natural to define a utility function over this space which determines how useful any model is independent of details of the model. Conversely, picking the required level of hazard mitigation allows one to specify a range of required precision and accuracy. Unlike the proverbial drunk looking for his keys under the light, model-builders can resist the temptation to build an accurate model at a useless precision solely because it seems possible. Furthermore, the gradient of the utility function can be estimated at the location of any model. Thus, given a model, one can say whether further efforts would be better spent on improving accuracy or precision.

This is not to suggest that determining the utility function is easy. In fact, considering the attempts at cost/benefit analyses in other areas where human life is at stake, it may even be impossible. Nonetheless, there may be enough agreement on its overall shape to guide research.

For example, consider the utility of the three forecasts introduced in the previous section. Forecast 1 is already part of the popular folklore, yet it seems to have had little effect on society as a whole. There is no massive migration away from seismically active areas. Its main effect has probably been in development of engineering standards for earthquake resistant structures. The utility function at this point in the space of models is fairly high but very flat. Tightening either the precision or the accuracy of the prediction will not make much difference in its effect on mitigation, and would thus not be worth the effort it would probably take. However, applying similar techniques to other areas, for example by searching for previously unknown faults under Los Angeles, would be worthwhile.

In contrast, the gradient of the utility function is quite large in the region of forecast 2. Depending on the time scale, one might be able to get individuals to "duck and cover"; get traffic off bridges and tunnels; or evacuate dangerous structures. Moreover, the actions one might take are more tightly constrained by the accuracy of the forecast. Issuing too many false positives ("crying wolf") will lead people to ignore the warnings. Thus the gradient is large in the directions of both precision and accuracy.

The most obvious mitigation strategy in the case of forecast 3 would be evacuation, as is done for hurricanes or volcanic eruptions. Clearly, the cost/benefit analysis in this example will depend crucially on the false positive rate and less so on the precision. Hence the utility function is fairly high and its gradient points mostly along the accuracy axis.

2.3 HOW GOOD IS A MODEL?

Once again, the question requires further definition. Models that try to predict discrete events (in this case, for example, large magnitude earthquakes) can be scored on the basis of false positives and false negatives. Models that try to predict entire time series are often scored on a mean-squared-error basis for convenience. A more relevant measure might be obtained from a cost-benefit analysis.

Practically speaking, there are few enough models which claim to offer predictability that the details of measuring performance are probably not yet important. More relevant is the question of how well a model generalizes from the past to the future. That is, though one model may fit past data much better than another, it may generalize more poorly and thus give poorer predictions than the other. This is a serious problem for models which claim to predict rare discrete events.

The reason is that, paradoxically, it is often too easy to find patterns in historical data which seem to be precursors for these events.

There are several approaches for addressing this question, including: controlling the complexity of models; cross-validating results on data that was not used for training; testing the model's robustness to small changes in its parameters; and testing the model's performance on random data. At the least, careful statistical significance testing of the model—including a clear statement of the null hypothesis of no precursors—are required.

3. WHAT DYNAMICS GIVE RISE TO USEFUL MODELS?

3.1 WHAT ARE PRECURSORS?

Most people probably think of precursors as seismic signals which occur shortly before a large event. For a data-driven model, though, precursors are any pattern in the data which can be used to predict events. Precursors could be on any time scale and involve any sort of spatiotemporal data. Thus, for example, the number of events above a certain magnitude in a certain region is a precursor if it predicts future activity. Indeed, the entire historical distribution of seismic activity is a precursor to the extent that it provides a statistical model of the probability of an event.[1]

3.2 HOW ARE DYNAMICS, PRECURSORS, AND MODELS RELATED?

Let us consider the dynamics that could underlie each of the example forecasts. Each of these dynamics suggests a different approach to modeling.

The first kind of prediction naturally arises from a stationary random process. The only relevant information (precursor) is the spatiotemporal distribution of events. Data-driven modeling consists of determining the best approximation to the historical distribution (in this case the Gutenberg-Richter scaling law).[5] Determining predictability is reduced to debating the validity of extrapolation to rare events and the assumption of stationarity. Of course, a bottom-up model can also produce this kind of prediction if it yields scaling laws. This kind of model might have the advantage of being able to determine the correct form of scaling laws *a priori* and thus determine the likelihood of rare events.

The second prediction is the sort one might make based on time series analysis of observations of a fairly simple noisy deterministic system. I will refer to this as "dripping faucet" dynamics. There is a collection of variables which constitutes the

---

[1]Under this interpretation, we must admit that at least one clear precursor, that leading to the Gutenberg-Richter scaling law, has been identified.

state of the system at any given time. The system's evolution is largely determined by its current state. The evolution function is usually reasonably smooth. The current state of the system can be deduced from the recent history of the time series, even if not all the variables that make up the state space can be observed directly. A model is built by embedding a set of observations in a vector space. Prediction then becomes nothing more than fitting a surface to the observations.[2,3] The precursors produced by dripping faucet dynamics are any set of observations that permit determination of the system's state. These are likely to include at least the recent history of seismic events in a nearby region.

The third prediction arises naturally from a complex system which contains a coherent structure. For example, consider the long-lived, spatially coherent structure called a hurricane. The precursors in this system are the existence, position, and velocity of the hurricane. There might be other, more subtle meteorological signals that a hurricane was about to develop, but it is not clear that these provide any *useful* information not contained in the obvious ones.[2] A traditional model might start with partial differential equations (PDEs) describing atmospheric dynamics. Guided by knowledge of the existence of coherent structures, it might be possible to reduce the PDEs to a small set of ordinary differential equations which control the coherent structure's dynamics. In contrast, without any knowledge of the underlying PDEs, a data-driven model can detect the structure and study its most relevant aspect—the space-time trajectory.

There is, of course, some overlap among the approaches suggested by these different kinds of dynamics. A time series model might be improved by taking into account the stationary distribution of events. The flows in state space which might be revealed by time series analysis could be considered a coherent structure which organizes the dynamics.

---

4. WHY HAVE WE SEEN NO CLEAR PRECURSORS?

As John Rundle has pointed out, there are only three possibilities:

- There aren't any.
- There are some in the data, but we haven't found them yet.
- There are some, but not in the data currently available.

The first possibility is the worst case, not only because it means modeling efforts are doomed, but also because it is inherently not provable, only falsifiable. As long as there are immense benefits to be gained from finding precursors, resources will be directed to the problem whether it is solvable or not.

---

[2] Useful in the sense of providing more timely warning for evacuation.

More interesting are reasons why the second and third possibilities might be true.

First, suppose there are precursors in currently available data which haven't yet been recognized. If the dynamics is based on large coherent structures, it can *only* be that we are not looking at the data in the right way. In particular, we must have overlooked the scale at which the coherent structure exists. For instance, we might have complete knowledge of the atmospheric pressure field of the planet, but not notice hurricanes because we look only at one square meter patches. This is an admittedly absurd analogy, but there could be a similar absurdity in the way we look at seismic data.

If, instead, the system exhibits dripping faucet dynamics and the precursors are in the available data, the problem must be that we have not searched hard enough for them. While this is undoubtedly a difficult and expensive search, the potential benefits clearly justify undertaking it as long as there seems to be a reasonable chance of success. In practice, this means at least trying new methods of data analysis as they become available.

Second, suppose the precursors are not in the currently available data. In the case of "coherent structure" dynamics, it as if we are trying to find hurricanes without satellite photos or shipping reports, or predict the weather without looking upstream. In this case, knowledge of the existence and scale of the coherent structure immediately tells us what additional observations are required. Clear understanding of the potential benefits of such a model can help justify the expense involved in collecting the data.

For dripping faucet dynamics, the situation could be much worse. The state space needed to adequately describe the dynamics could be high-dimensional (i.e., the system might have many degrees of freedom), which would place fundamental constraints on the amount of data needed to build a model (exponential in the dimension). The dynamics could be not only nonlinear, but also chaotic, which would imply that only limited-term forecasts could be made and might place such severe restrictions on required measurement precision that the models would be useless for hazard mitigation. Even worse, the dynamics could be both high-dimensional and chaotic. The best that could be hoped for in this case would be estimates of the fundamental limits to predictability.

5. CONCLUSIONS

Data-driven models of seismic activity provide a complementary approach to traditional models based on fundamental physics. These models may provide more useful predictions than fundamental models for a complex system because they are, by design, geared toward prediction of specific events.

I have tried to indicate the kinds of dynamics which give rise to various kinds of precursors and the types of models that capture those dynamics well. The fact that no clear precursors have been identified yet may place constraints on the possible kinds of dynamics and the inherent limitations of predictive models. These, in turn, may make it possible to determine the feasibility of hazard mitigation even in the absence of a model. Failing that, the limits may help prioritize questions which need to be resolved.

ACKNOWLEDGEMENTS

Thanks to the organizers of this workshop for inviting my comments and to the Santa Fe Institute for providing me with the facilities to get these ideas into a reasonably coherent form.

REFERENCES

1. Bak, P., C. Tang, and K. Wiesenfeld. "Self-Organized Criticality." *Phys. Rev. A* **38** (1988): 364–374.
2. Casdagli, M., and S. Eubank, eds. *Nonlinear Modeling and Forecasting.* SFI Studies in the Sciences of Complexity, Proc. Vol. XII. Reading, MA: Addison-Wesley, 1992.
3. Casdagli, M., D. Des Jardins, S. Eubank, J. D. Farmer, J. Gibson, N. Hunter, and J. Theiler. "Nonlinear Modeling of Chaotic Time Series: Theory and Applications." In *Applied Chaos*, edited by J. H. Kim and J. Stringer, 335–380. New York: John Wiley & Sons, 1992.
4. Gabrielov, Andrei, and William I. Newman. "Seismicity Modeling and Earthquake Prediction: A Review." 1993.
5. Gutenberg, M., and C. F. Richter. "Frequency of Earthquakes in California." *Bull. Seism. Soc. Am.* **34** (1944): 185–188.
6. Keilis-Borok, V. I., and I. M. Rotwain. "Diagnosis of Time of Increased Probability of Strong Earthquakes in Different Regions of the World: Algorithm CN." *Phys. Earth & Plant. Int.* **61** (1990): 57–72.
7. Shaw, R. S. *The Dripping Faucet as a Model Dynamical System.* Santa Cruz, CA: Aerial Press, 1984.

C. G. Sammis,* D. Sornette, and H. Saleur†**

*Department of Earth Sciences, University of Southern California, Los Angeles, CA 90089-0740

**Laboratoire de Physique de la Matière Condensée, CNRS, URA 190 Université des Sciences, B. P. 70, Parc Valrose, 06108 Nice Cedex 2, France

†Department of Physics and Astronomy, University of Southern California, Los Angeles, CA 90089-0484

Complexity and Earthquake Forecasting

There is mounting evidence that earthquake prediction based on seismicity is a problem involving the collective behavior of a spatially extended regional fault network. Many large earthquakes, for example, are preceded by an increase in the number of intermediate-sized events within a large region surrounding the mainshock. This concentration of larger events toward the end of the seismic cycle produces a corresponding increase in energy release (or any other equivalent measure of the deformation) which has recently been fit by Bufe and coworkers to a power-law time-to-failure equation which predicts the time of the mainshock. In a recent paper, Sornette and Sammis[33] show that the power-law time-to-failure equation can be derived from renormalization group theory where the mainshock is viewed as a critical point. The advantage of this treatment is that we are able to also derive the first-order correction to scaling. For the case of a discrete scale-invariant system, this correction takes the form of a periodic function of the logarithm of the time to failure which modulates the power-law increase. We show that this log-periodic correction is evident in the regional seismicity data which preceded the 1989 Loma Prieta earthquake[6] and the current buildup of activity on several segment of the Aleutian arc.[7] Identification of these log-periodic fluctuations allows a significantly more accurate prediction of the time of the mainshock to be made at a significantly earlier time in the sequence than does the simple power law.

INTRODUCTION

"Only charlatans and fools predict earthquakes." This quote, ascribed to Charles Richter (although apparently never published), is a fairly accurate summation of the state of earthquake prediction in the early 1960s. At that time there was no physical basis upon which to make a prediction of the time, place, or magnitude of a future event. However, two discoveries in the mid-1960s to late 1960s suggested that earthquake prediction might be a suitable subject for serious scientific research. These were the development of the plate tectonics model and the observation of physical precursory events which immediately preceded the failure of rock specimens in the laboratory.

The development of the plate tectonics model in the mid-1960s explained not just the observed spatial pattern of seismicity, but also allowed an estimate of the average rate of that seismicity at specific segments of the plate boundaries. For example, if the long-term average rate of motion on a plate boundary is divided by the average slip of the largest "characteristic earthquakes" on a given segment, then the average recurrence interval of the characteristic event may be calculated. For the San Andreas fault in southern California, this calculation gives a recurrence interval of about 140 years. The fact that this same recurrence interval can also be calculated by fitting the shorter recurrence intervals of smaller events scattered across southern California to the Gutenberg-Richter relation $\log(N_T) = a - bM$ (where N_T is the total number of events of magnitude M or greater and a and b are empirical constants) and extrapolating to $M = 8$ suggests that the largest characteristic regional events may be mechanistically related to the more broadly dispersed seismicity in that region.[16]

This average recurrence interval predicted by both plate tectonics and regional seismicity rates was confirmed by Sieh et al.[32] using paleoseismic excavations on the San Andreas fault near Palmdale, CA. The average recurrence interval of the last ten large events which ruptured this segment of the San Andreas since about A.D. 500 is about 132 years. Since the last large event occurred in 1857, one might have expected the next near 1989. However, the large event that preceded the 1857 event occurred in 1812 giving a recurrence interval of only 45 years. In fact Sieh et al. found that the earthquakes occurred in four clusters of two or three events having recurrence intervals of significantly less than 100 years between events, while the intervals between clusters were significantly longer lasting between 200 and 332 years. This is not good news for earthquake forecasting. Neither is a similar clustering in time in the Parkfield region[5] where the seemingly regular recurrence interval of about 20 years between the last six $M = 6$ events has already extended to almost 30 years since the last event in 1966 resulting in the prediction for that region to formally be declared a failure. From the perspective of insight gained from recent advances in nonlinear mechanics, such observation have led many to conclude that earthquakes, like the weather, are the output of a sensitive nonlinear system, and therefore are inherently unpredictable.

Even if earthquakes are unpredictable in a mechanistic sense, the hope has been that they give some warning in the form of physical precursors. Indeed, careful observation of rock failure in the laboratory has revealed that, beginning at stress levels about halfway to failure, dilatant microcracks nucleate and grow. Near failure, this microcracking is so extensive that the sample volume actually increases—a phenomenon termed "dilatancy" in reference to a similar phenomenon which precedes failure in granular materials. Dilatancy has many observable consequences in the laboratory which, if they are observed in the field, should provide measurable precursors to earthquakes. Elastic wave velocities, electrical resistivity, groundwater levels, ground water chemistry, elevation and local gravity should all show characteristic signatures which are predicted by the "dilatancy-diffusion model" of earthquakes developed by Nur[22] and expanded upon by Whitcomb et al.[37] and Scholz et al.[31] Unfortunately, as Turcotte[35] points out in his review of recent progress in earthquake prediction, the physical precursors observed to precede failure in the laboratory or expected as a consequence of analogous distributed fracturing and the associated movement of ground water in the crust have not been consistently observed in the field.

More recently, a third approach to earthquake forecasting has emerged based on the observation that many large earthquakes have been preceded by an increase in the number of intermediate-sized events.[6,7,14,17,34] The relation between these intermediate-sized events and the subsequent main event has only recently been recognized because the precursory events occur over such a large area that they do not fit prior definitions of foreshocks.[10] They do, however, make sense in the context of nonlinear systems where long-range correlations are expected to precede critical phenomena. Previous investigators[6,7,34] have attempted to quantify this increase in activity at the end of a seismic cycle to predict the time of the mainshock. In this chapter, we extend these analyses using the technique of renormalization group (RG) theory to model temporal patterns of seismicity which precede large events. Preliminary results, discussed below, look promising.[30,33]

OBSERVATIONS OF INTERMEDIATE-SIZED PRECURSORS TO GREAT EVENTS

There is mounting evidence that the rate of occurrence of intermediate earthquakes increases in the tens of years preceding a major event. Sykes and Jaumé[34] present evidence that the occurrence of events in the range $M = 5.0 - 5.9$ accelerated in the tens of years preceding the large San Francisco Bay area quakes in 1989, 1906, and

1868 and the Desert Hot Springs earthquake in 1948. Lindh[17] points out references to similar increases in intermediate seismicity before the large 1857 earthquake in southern California and before the 1707 Kwanto and the 1923 Tokyo earthquakes in Japan. More recently, Jones[11] has documented a similar increase in intermediate activity over the past eight years in southern California. However, this increase in activity has been limited to events in excess of $M = 5.0$; no increase in activity is apparent when all events $M \geq 4.0$ are considered. Similarly, Ellsworth et al.[9] reported that the increase in activity was also limited to events larger than $M = 5$ prior to the 1989 Loma Prieta earthquake in the San Francisco Bay area. Raleigh et al.[23] point out that the intermediate precursors occur in the general vicinity of the coming rupture, but usually not along the rupture itself. They note that the first observation of this general phenomenon was by Mogi[19] who found that moderate to large earthquakes became more numerous in a broad region surrounding the rupture zones of great Japanese earthquakes within a few decades of those great events. Because these foreshocks tended to occur in the region surrounding the impending shock, but not on the fault which produced it, this pattern has been termed the "Mogi donut." Bufe and Varnes[6] analyzed the increases in activity which preceded the 1989 Loma Prieta earthquake in the San Francisco Bay area while Bufe et al.[7] have documented a current increase in seismicity in several segments of the Aleutian arc and forecast a large event on one or more segments by about 1996.

QUANTIFYING OBSERVED INCREASES IN REGIONAL SEISMICITY

When viewed in the context of the seismic cycle, this increase in intermediate-sized events dominates the regional energy release rate producing a rapid increase toward the end of the cycle as illustrated in Figure 1 from Ellsworth et al.[9] There have been several attempts to quantify this regional increase in energy release and thereby to predict future large earthquakes in the area. Based on the four events which they studied, Sykes and Jaumé[34] suggest a criterion based on the cumulative regional moment released, ΣM_0. They observed that the mainshocks tended to occur within a few years of the time that ΣM_0 reached 6×10^{18} Nm. They also attempted to fit the final rapid increase in moment release to the empirical equation $\ddot{\Omega} - A\dot{\Omega}^\alpha = 0$ which was shown by Voight[36] to describe rate-dependent failure in a number of materials. For $1 < \alpha < 2$, Ω becomes very large at a finite time and failure may be predicted. For most materials, α is slightly less than 2. However, Sykes and Jaumé were not successful in using Voight's equation with $\Omega = \Sigma M_0$ to estimate the time of the mainshocks from their data.

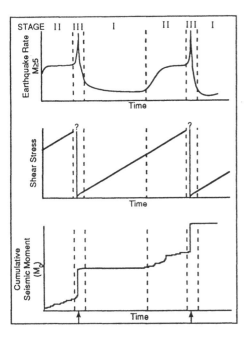

FIGURE 1   A schematic diagram of the hypothesized stages of the seismic cycle from Ellsworth et al.[9] Note the increase in cumulative seismic moment which precedes failure in the lower graph.

More recently, Bufe and Varnes[6] fit the cumulative "Benioff strain" ($\Sigma E^{1/2}$) which preceded the 1989 Loma Prieta earthquake to a power-law relation of the form

$$\Sigma E^{1/2} = A + B(t_f - t)^m \qquad (1)$$

and obtained a useful prediction of the time, t_f, and the energy $(A - \Sigma E^{1/2})^2$ of the mainshock. Bufe et al.[7] applied the same analysis to the increase in seismic activity which is currently occurring on several segments of the Aleutian arc to predict one or more $M = 7$ events by about 1996.

PHYSICAL MECHANISMS FOR PRECURSORY REGIONAL SEISMICITY

Why should the energy (or moment) release increase as a power law of the time to failure? Bufe and Varnes[6] justify this behavior in terms of either runaway crack propagation or damage mechanics, both of which yield a power-law failure equation of the form in Eq. (1). It is simple and informative to show that the most elementary form of damage mechanics yields Eq. (1). If V_i is the volume fraction of intact rock at time t and V_f is the volume fraction which has failed such that $V_i + V_f = 1$, then the average local stress, σ_{local}, will be related to the remote regional stress σ_{remote}

by the average area fraction of intact rock which is carrying the load as $\sigma_{local} = \sigma_{remote}/A_i = \sigma_{remote}/V_i$ (since the average cross-section area fraction equals the average volume fraction). If the average failure rate is proportional to the average local stress raised to some power β, we have

$$\frac{dV_f}{dt} = -\frac{dV_i}{dt} = (\text{constant})(\sigma_{local})^\beta = (\text{constant})\left(\frac{\sigma_{remote}}{V_i}\right)^\beta \tag{2}$$

which can be integrated

$$\int_t^{t_f} dt = -(\text{constant})\int_{V_i}^0 V_i^\beta dV_i \tag{3}$$

to yield

$$t_f - t = (\text{constant})V_i^{\beta+1}. \tag{4}$$

If we assume that the energy release is proportional to the failed volume, Eq. (4) may be written

$$E(t) = c_1 V_f(t) = c_1[1 - V_i(t)] = c_1[1 - c_2(t_f - t)^{1/(\beta+1)}] \tag{5}$$

where $E(t)$ is the cumulative energy released by the volume fraction which has failed up to time t, and c_1 and c_2 are constants. Similarly, if we assume that the Benioff strain associated with each event is proportional to the increment in $E^{1/2}$, we also expect the cumulative Benioff strain to vary as a power of the time to failure as in Eq. (1).

However, Sornette and Sammis[33] point out that a power-law increase in regional energy release can also be expected if the progressive failure of a heterogeneous region is viewed as being analogous to a phase transition. The previous treatment is then seen to be a mean field approximation of the full renormalization group theory to be developed below. In this view, a large regional event is considered a critical point, and the increase in regional energy release associated with the increase in intermediate events is due to the progressive increase in correlation length which develops in the scaling regime close to the critical point. Sornette and Sammis[33] use renormalization group (RG) theory to derive Eq. (1). The basic RG equations are

$$x' = \phi(x), \tag{6}$$

$$F(x) = g(x) + \frac{1}{\mu}F(\phi(x)), \tag{7}$$

where $x = t_f - t$, t_f is the time of global (regional) failure for the region under consideration, ϕ is called the RG flow map, $F(x) = \varepsilon(t_f) - \varepsilon(t)$ such that $F = 0$ at the critical point, and μ is a constant describing the scaling of the seismic release rate upon the discrete time rescaling (6). The function $g(x)$ represents the non-singular part of the function $F(x)$. We assume as usual that the function $F(x)$ is continuous and that $\phi(x)$ is differentiable. If $x = 0$ denotes a fixed point ($\phi(0) = 0$) and $\phi(x) = \lambda x + \ldots$ is the corresponding linearized transformation, then the solution of Eq. (7) close to $x = 0$ is given by Eq. (1), i.e. $F(x) \sim x^{\alpha}$, with $\alpha = \mathrm{Log}\mu/\mathrm{Log}\lambda$.

More significantly, for the case of *discrete* renormalizations as in Eqs. (6)–(7), Eq. (1) is not the only solution to the RG equations. The power-law solution (1) multiplied by a periodic function p of $\log(t_f - t)$ having period $\log\mu$ of the form

$$F(x) = F_0(x)p(\log F_0(x)) \tag{8}$$

is also a solution where $F_o(x)$ is a special solution. This periodic (in $\mathrm{Log}x$) correction to the dominant scaling of Eq. (1) amounts to considering a complex critical exponent α, since $\mathrm{Re}[x^{\alpha'+i\alpha''}] = x^{\alpha'}\cos(\alpha''\mathrm{Log}x)$.

Although previous papers have explored the analogy between brittle failure and phase transitions[25,26,27] and have applied RG techniques to the failure and conductivity in rock,[1,18] none have recognized the possibility of these log-periodic fluctuations. This is probably because this is not an allowed solution for the continuously renormalizable variables usually encountered in thermodynamic problems. However, log-periodic fluctuations have been observed in physics problems which involve a preexisting heterogeneous structure such as the superconducting transition on a Sierpinski gasket fractal network,[8] the failure[30] of a hierarchical fiber bundle model studied by Newman et al.,[21] and the acoustic emissions preceding failure of a fiber-reinforced composite material.[2] It is not clear at this point whether a preexisting hierarchical structure is required (such as the Sierpinski gasket or fiber bundle), or whether heterogeneous systems can "self-organize" their own hierarchy.[30]

Sornette and Sammis[33] fit a log-periodic function of the form

$$\varepsilon(t) = A + B(t_f - t)^m \left[1 + C\cos\left(2\pi\frac{\log(t_f - t)}{\log\lambda} + \psi \right) \right], \tag{9}$$

to the cumulative Benioff strain preceding the 1989 Loma Prieta earthquake originally fit to the power-law Eq. (1) by Bufe and Varnes.[6] Figure 2 shows both fits. Note that the fluctuations about the power law are well fit by the additional log-periodic factor. The parameters and their uncertainties (as determined using the Levenberg-Macquardt nonlinear fitting algorithm) are summarized in Table 1. Note that even though there are more parameters in the new fit (Eq. (9)), the time of failure t_f is better constrained due to the additional structure in the curve. In fact, as illustrated in Figure 3, the additional structure makes it possible to make a more accurate prediction significantly earlier in the sequence. Figures 4 and 5 shows this same analysis for the recent increase in seismicity documented by Bufe et al.[7] for the Kommandorski Island segment of the Aleutian arc. The parameters of the power-law Eq. (1) and corrected Eq. (9) fit are given in Table 1. Again, the additional structure allows a more precise prediction to be made much earlier in the sequence.

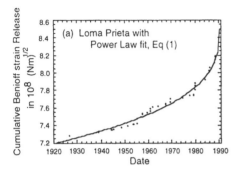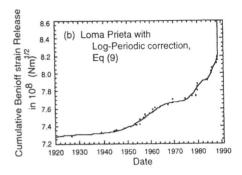

FIGURE 2   Cumulative Benioff strain released by magnitude 5 and greater earthquakes in the San Francisco Bay area prior to the 1989 Loma Prieta earthquake (from Bufe and Varnes, 1993). In (a) the data have been fit to the power-law Eq. (1) as in Bufe and Varnes. In (b), the data have been fit to Eq. (9) which includes the first-order correction to scaling. Parameters of both fits are given in Table 1.

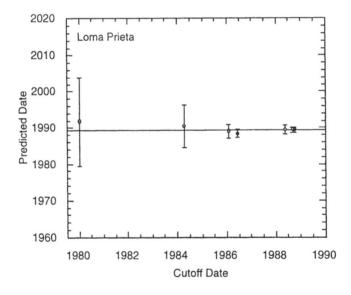

FIGURE 3   Predicted date of the Loma Prieta earthquake using Eq. (9) to fit all the data prior to the date shown on the abscissa. Not that a useful prediction is obtained after 1980, and that the quality of the prediction tends to improve as the date of the earthquake is approached. The horizontal line is the actual date of the Loma Prieta earthquake (October 17, 1989).

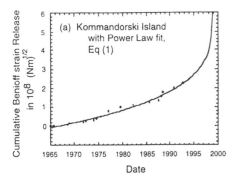 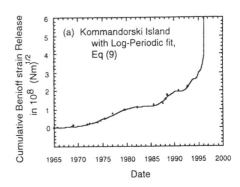

FIGURE 4   Cumulative Benioff strain released by magnitude 5.2 and greater earthquakes in the Kommandorski Island segment of the Aleutian Island seismic zone.[7] In the left frame (a) the data have been fit to the power-law Eq. (1) as in Bufe et al. except that we did not fix the exponent at $m = 0.3$ as in that paper. In the right frame (b) the data have been fit to Eq. (9) which includes the first-order correction to scaling. Parameters of both fits are given in Table 1.

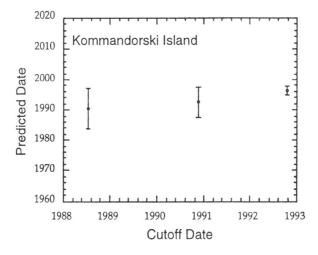

FIGURE 5   Predicted date of an impending earthquake in the Kommandorski segment when all the data up to the date shown on the abscissa are fit to Eq. (9).

TABLE 1 Parameters found by fitting time-to-failure equations to the cumulative Benioff strain.

Power-law fit Equation (1)	Loma Prieta	Kommandorski Island
A	8.50 ± 0.73	6.23 ± 26.9
B	-0.29 ± 0.44	-2.49 ± 19.3
m	0.35 ± 0.23	0.26 ± 1.0
t_f	1990.3 ± 4.1	1998.8 ± 19.7
Corrected fit Equation (9)		
A	8.46 ± 0.24	4.95 ± 2.25
B	-0.30 ± 0.16	-1.88 ± 1.77
m	0.34 ± 0.08	0.28 ± 0.14
C	-0.050 ± 0.015	0.040 ± 0.023
λ	3.13 ± 0.14	2.50 ± 0.26
ψ	1.45 ± 0.89	-3.25 ± 2.52
t_f	1989.3 ± 0.8	1996.3 ± 1.1

DISCUSSION

These observations immediately raise a number of questions. First, how universal are these log-periodic bursts of seismicity? Mogi[20] pointed out that a rich regional foreshock sequence is to be expected if the material is mechanically heterogeneous. Hence such temporal fluctuations may be limited to regions which have a complex array of faults which span a wide range of scales such as the San Francisco Bay area and the big-bend region in southern California. They might not be expected in simpler tectonic settings where motion is accommodated on one well-aligned structure.

Second, are these temporal sequences limited to the largest events in a region, or are the intermediate-sized precursors expected to have their own precursory sequence developed over a shorter time scale and smaller region? In this view, each earthquake represents the failure of some surrounding region the size of which scales with the size of the event. Each earthquake is, at the same time, the failure of its local region and a part of the precursory failure sequence of an even larger event. Failure of the crust can be thought of as a scaling-up process in which failure at one scale is part of the damage accumulation pattern at a larger scale.[15] The hierarchical fiber bundle model[21] provides an excellent simple mechanical analog for this scaling-up process . Whenever a fiber fails at any scale, its load is carried by the remaining intact fibers thus leading to a power-law increase in energy release near failure as discussed above. However, unlike the simple damage model discussed above, stress is redistributed only to neighboring fibers in the same bundle. Hence each bundle has its own precursory failure sequence but is also part of the failure sequence of the larger bundle in which it is an element. In the geological context, each bundle represents a subregion of the hierarchical fault network. This view is consistent with observations that seismicity forms fractal clusters in both space and time,[12,13,24] although the nested hierarchical structure implied here has not been established.

Third, why analyze the Benioff strain? We chose this measure of deformation because it was fit to the power-law time-to-failure equation by Bufe and co-workers and we wished to directly contrast the RG results with their work. However, why not renormalize the energy release itself or, for that matter, any other power q in the expression

$$\varepsilon_q(N) = \sum_{n=1}^{N} E_n^q. \tag{10}$$

Working with the energy has the advantage that it is dominated by the largest events since each unit decrease in magnitude corresponds to about ten times as many events but only about one-thirtieth of the energy for each. This situation reverses at $q \sim 2/3$ where each magnitude releases roughly the same energy. For $q < 2/3$ progressively smaller events make an increasingly large contribution to the sum of Eq. (10). It may be that the Benioff strain ($q = 1/2$) is a compromise between $q = 1$ which is dominated by a few events and thus has a large stochastic uncertainty and small q which does not give special weight to the larger events observed to preceed great earthquakes, and which introduces a large uncertainty associated with gaps in the seismic catalogs for smaller events.

Finally, what constraints, in addition to structural complexity are required to produce the observed fluctuations? Are they the result of a discrete structural hierarchy in the underlying fault network analogous to the Sierpinski gasket or fiber bundles. Or does the system self-organize its own hierarchical structure as recently suggested by Saleur et al.[30] Whatever the underlying physics, if these log-periodic fluctuations are observed, they offer the possibility of improving the precision and lead-time of intermediate term earthquake forecasts.

REFERENCES

1. Allegre, C. J., J. L. Le Mouel, and A. Provost. "Scaling Rules in Rock Fracture and Possible Implications for Earthquake Prediction." *Nature* **297** (1982): 47–49.
2. Anifrani, J. C., C. Le Floc'h, D. Sornette, and B. Souillard. "Universal Log-Periodic Correction to Renormalization Group Scaling for Rupture Stress Prediction from Acoustic Emissions." *J. Phys. I France* (1995): 631–638.
3. Bak, P., and C. Tang. "Earthquakes as a Self-Organized Critical Phenomenon." *J. Geophys. Res.* **94** (1989): 15635–15637.
4. Bak, P., C. Tang, and K. Wiesenfeld. "Self-Organized Criticality." *Phys. Rev. A* **38** (1989): 364–374.
5. Bakun, W. H., and T. V. McEvilly. "Recurrence Models and Parkfield, California Earthquakes." *J. Geophys. Res.* **89** (1984): 3051–3058.
6. Bufe, C. G., and D. J. Varnes. "Predictive Modeling of the Seismic Cycle of the Greater San Francisco Bay Region." *J. Geophys. Res.* **98** (1993): 9871–9883.
7. Bufe, C. G., S. P. Nishenko, and D. J. Varnes. "Seismicity Trends and Potential for Large Earthquakes in the Alaska-Aleutian Region." *PAGEOPH* **142** (1994): 83–99.
8. Doucot, B., W. Wang, J. Chaussy, B. Pannetier, and R. Rammal. "First Observation of the Universal Periodic Corrections to Scaling; Magnetoresistance of Normal-Metal Self-similar Networks." *Phys. Rev. Lett.* **57** (1986): 1235–1238.
9. Ellsworth, W. L., A. G. Lindh, W. H. Prescott, and D. J. Herd. "The 1906 San Francisco Earthquake and the Seismic Cycle." *Am. Geophys. Union, Maurice Ewing Monogr.* **4** (1981): 126–140.
10. Jones, L. M., and P. Molnar. "Some Characteristics of Foreshocks and Their Possible Relationship to Earthquake Prediction and Premonitory Slip on Faults." *J. Geophys. Res.* **84** (1979): 3596–3608.
11. Jones, L. M. "Foreshocks, Aftershocks, and Earthquake Probabilities: Accounting for the Landers Earthquake." *Bull. Seism. Soc. Am.* **84** (1994): 892–899.
12. Kagan, Y. Y. "Spatial Distribution of Earthquakes: The Three-Point Moment Function." *Geophys. J. R. Astron. Soc.* **67** (1981): 697–717.
13. Kagan, Y. Y. "Spatial Distribution of Earthquakes: The Four-Point Moment Function." *Geophys. J. R. Astron. Soc.* **67** (1981): 719–733.
14. Keilis-Borok, V. I., L. Knopoff, I. M. Rotwain, and C. R. Allen. "Intermediate-Term Prediction of Times of Occurrence of Strong Earthquakes in California and Nevada." *Nature* **335** (1988): 690–694.
15. King, G. C. P. "The Accommodation of Large Strains in the Upper Lithosphere of the Earth and Other Solids by Self-similar Fault Systems: The Geometrical Origin of b-Value." *PAGEOPH* **121** (1983): 761–815.

16. King, G. C. P., and C. G. Sammis. "The Mechanisms of Finite Brittle Strain." *PAGEOPH* **138** (1992): 611–640.
17. Lindh, A. G. "The Seismic Cycle Pursued." *Nature* **348** (1990): 580–581.
18. Madden, T. R. "Microcrack Connectivity in Rocks: A Renormalization Group Approach to the Critical Phenomena of Conduction and Failure in Crystalline Rocks." *J. Geophys. Res.* **88** (1983): 585–592.
19. Mogi, K. "Some Features of Recent Seismic Activity in and near Japan 2: Activity Before and After Great Earthquakes." *Bull. Eq. Res. Inst. Tokyo Univ.* **47** (1969): 395-417.
20. Mogi, K. "Rock Fracture and Earthquake Prediction." *J. Soc. Mater. Sci., Jpn.* **23** (1974): 320–329 [in Japanese].
21. Newman, W., A. Gabrielow, T. Durand, S. L. Phoenix, and D. Turcotte. "An Exact Renormalization Model for Earthquakes and Material Failure, Statics and Dynamics." *Physica D* **77** (1994): 200–216.
22. Nur, A. "Dilatancy, Pore Fluids, and Premonitory Variations in ts/tp Travel Times." *Bull. Seism. Soc. Am.* **62** (1972): 1217–1222.
23. Raleigh, C. B., K. Sieh, L. R. Sykes, and D. L. Anderson. "Forecasting Southern California Earthquakes." *Science* **217** (1982): 1097–1104.
24. Robertson, M. C., C. G. Sammis, M. Sahimi, and A. Martin. "Fractal Analysis of Three-Dimensional Spatial Distributions of Earthquakes with a Percolation Interpretation." *J. Geophys. Res.* **100** (1995): 609–620.
25. Rundle, J. B. "A Physical Model for Earthquakes, 1. Fluctuation and Interaction." *J. Geophys. Res.* **93** (1988): 6237–6254.
26. Rundle, J. B. "A Physical Model for Earthquakes, 2. Application to Southern California." *J. Geophys. Res.* **93** (1988): 6255–6274.
27. Rundle, J. B. "A Physical Model for Earthquakes, 3. Thermo-dynamical Approach and Its Relation to Nonclassical Theories of Nucleation." *J. Geophys. Res.* **94** (1989): 2839–2855.
28. Sahimi, M., and S. Arbabi. "Mechanics of Disordered Solids, 2. Percolation on Elastic Networks with Bond-Bending Forces." *Phys Rev. B* **47** (1993): 703–712.
29. Sahimi, M., and S. Arbabi "Mechanics of Disordered Solids, 3. Fracture Properties." *Phys Rev. B* **47** (1993b): 713–722.
30. Saleur, H., C. G. Sammis, and D. Sornette. "Discrete Scale Invariance, Complex Fractal Dimensions, and Log-Periodic Fluctuations in Seismicity." *J. Geophys. Res.* (1995): submitted.
31. Scholz, C. H., L. R. Sykes, and Y. P. Aggarwal. "Earthquake Prediction: A Physical Basis." *Science* **181** (1973): 803–809.
32. Sieh, K., M. Stuiver, and D. Brillinger. "A More Precise Chronology of Earthquakes Produced by the San Andreas Fault in Southern California." *J. Geophys. Res.* **94** (1989): 603–623.
33. Sornette, D., and C. G. Sammis. "Complex Critical Exponents from Renormalization Group Theory of Earthquakes: Implications for Earthquake Predictions." *J. Phys. I France* (1995): 607–619.

34. Sykes, L. R., and S. C. Jaumé. "Seismic Activity on Neighboring Faults as a Long-term Precursor to Large Earthquakes in the San Francisco Bay Area." *Nature* **348** (1990): 595–599.

35. Turcotte, D. L. "Earthquake Prediction." *Ann. Rev. Earth Planet. Sci.* **19** (1991): 263–281.

36. Voight, B. "A Relation to Describe Rate-Dependent Material Failure." *Science* **243** (1989): 200–203.

37. Whitcomb, J. H, J. D. Garmony, and D. L. Anderson. "Earthquake Prediction: Variation of Seismic Velocities Before the San Fernando Earthquake." *Science* **180** (1973): 632–641.

Susanna Gross
Department of Geological Sciences and Cooperative Institute for Research in Environmental Sciences, University of Colorado, Boulder, CO 80309

On the Scaling of Average Slip with Rupture Dimensions of Earthquakes

In this chapter some consequences of heterogeneous fault rupture for the scaling properties of seismicity will be explored. A simple model of fault rupture is constructed in which fault slip is concentrated in small regions called sticking points. The sticking points have a power-law strength distribution and are scattered about the fault plane where the rupture forms. In this model, rupture begins at a randomly selected sticking point, and proceeds in four directions away from the hypocenter. Each rupture front breaks weaker sticking points it encounters, but may be stopped by stronger ones or by the fault edges. Sticking points weak enough to be broken contribute their energy to the rupture front, making the rupture stronger and able to break larger sticking points. The final moment and dimensions of the synthetic ruptures may be compared to natural earthquakes, and are found to have similar scaling properties. The slope and scatter of the observed relationship between seismic moment and rupture area are reproduced, as is a typical Gutenberg-Richter magnitude frequency distribution. When the average slip of these simulated ruptures is plotted versus length, ruptures significantly longer than the fault width exhibit slip that increases with rupture length as the square root of the length. This scaling is intermediate between the relationships proposed by Scholz[10] and Romanowicz and Rundle.[9]

Reduction & Predictability of Natural Disasters, Eds. Rundle, Turcotte, & Klein,
SFI Studies in the Sciences of Complexity, Vol. XXV, Addison-Wesley, 1996

INTRODUCTION
PREVIOUS WORK

The ideas discussed below have been developed in various forms by a large number of seismologists. The concept that most moment release in large earthquakes is concentrated into a few relatively strong points is quite old.[5] These strong points are generally called asperities, and may be responsible for the concentrations of slip that have been imaged in many inversions for the distribution of fault slip.[2,6,13] Fractal distributions of fault strength and/or stress have been used by von Seggern[12] to discuss implications for earthquake prediction and Huang and Turcotte[3] to explore variations in b-value through an earthquake cycle. Randomly distributed sub-events very analogous to the sticking points discussed here have been used to model seismic ground motion spectra.[16] The idea that strong points may stop fault rupture should be credited to Aki,[1] and in that role they are called barriers. Generally, fault rupture heterogeneity has not been much explored for its implications for earthquake scaling or magnitude-frequency statistics, and asperities and barriers have sometimes been seen as opposing concepts in the dynamics of fault rupture.

The scientific debate about the relationship between rupture length and average slip has been fueled by the practical importance of the problem, the poor quality of the data available, and the absence of a physical explanation for the observations.[11] The scaling relationship between rupture length and average slip for large earthquakes is of practical importance, because it governs how damaging the most destructive earthquakes on a fault of a given size may be. This particular relationship is difficult to assess on a purely observational basis, because the number of earthquakes decreases by roughly a factor of ten with every unit increase in magnitude. The number of events large enough so that W is limited by the thickness of the brittle crust is especially meager.

Both rupture length and average slip are difficult to determine accurately, which explains a portion of the scatter on Figure 1. The data on the figure were transformed from that presented by Romanowicz[8] with the assumption that all events had a constant rupture width of 20 km and an elastic modulus $\mu = 3 \times 10^{10}$ Pa. The low average slips of events shorter than 50 km could indicate that these events typically have rupture widths smaller than 20 km. Certainly, the average rupture lengths of events with low average slip are not well represented on Figure 1 because of the elimination of all events having rupture lengths below 20 km. There is a trend of increasing average slip with length visible in the figure, in spite of the sparseness and noisiness of the data, but no theories have been developed that fully explain this observation.

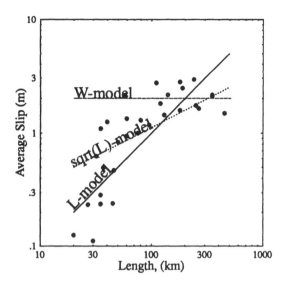

FIGURE 1 Average slip plotted versus length, for strike-slip faults, with lines representing various alternative scaling models.

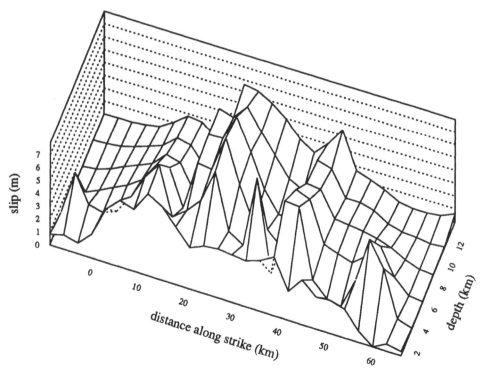

FIGURE 2 A model of the spatial distribution of slip during the 1992 Landers, California earthquake illustrates irregularity of slip in large earthquakes.

SIMPLE RUPTURE MODEL

The assumption that rupture is extremely heterogeneous is motivated in part by inversions for the distribution of slip which show great irregularity such as that in Figure 2. The slip inversion in the figure is unusually well constrained, but its features are similar to many others that have been determined in recent years. This model was developed by Wald and Heaton,[13] and is constrained by geodetic surveys, near-field and regional strong ground motion seismic records, broadband teleseismic waveforms, and measurements of surface offset. Along-strike measurements increase northward from the epicenter. Some of the features are as small as the resolution elements in the model. Irregularity of the slip distribution is not predicted by elasticity theory applied to uniform elastic solids, but may be explained by irregularity of material properties, geometry, or history.

The simple rupture model is designed to capture the essence of the physical processes involved in earthquake rupture while ignoring much of the complexity which exists in the real world. For example, there is no time scale explicitly included in this model. Ruptures begin at a hypocenter and run outward, but their speed is not defined, and the rupture dynamics is reduced to a set of simple rules. The fault is modeled as a flat two-dimensional surface with a bounded width and unbounded length. Sticking points are distributed uniformly and randomly upon the surface. These points could be regarded as points of contact between two irregular surfaces, and they could be regarded as stress concentrations, areas that have not slipped recently and therefore contain much of the energy available to drive ruptures. The size of the sticking points is also not explicit, but it is not great enough to place constraints upon their location. Sticking points have a power-law distribution of moment with frequency of occurrence, analogous to a Gutenberg-Richter distribution of earthquakes with magnitude and a b-value of 1.

Rupture begins at a randomly selected sticking point and moves outward in four directions, so there are exactly four rupture fronts for each earthquake. The fault is divided up into 90° quadrants, so there is no difference between rupture along strike and rupture along dip until a fault edge is reached. The earthquake continues to grow as long as any one of the four independent ruptures progresses. Each rupture front has a moment equal to the total moment content of all the sticking points which have been broken by that rupture front, and each of the four fronts starts with an equal share (1/4) of the moment of the first sticking point. Rupture on a front ceases when the front encounters either a sticking point stronger than the total moment content of the front, or the edge of the fault. The final rupture area is defined as the total distance ruptured toward the width of the fault multiplied by the total distance ruptured along the length of the fault. Average slip may be computed from the moment content of all sticking points broken and the area.

The synthetic ruptures in this model have a distribution of number N with magnitude greater than M_w similar to that observed in natural earthquakes and expressed in the Gutenberg-Richter relation:

$$\log_{10} N = A - bM_w.$$

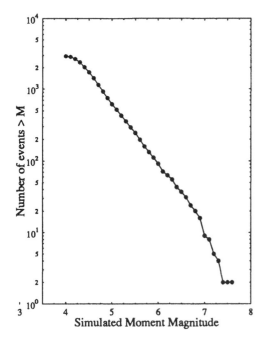

FIGURE 3 A distribution of simulated events with magnitude resembles a power law with b-value of 1.

An example of a simulated magnitude-frequency distribution is shown in Figure 3. Above magnitude 7.5, there is a increase in b-value, but this is due to ruptures breaking beyond the ends of a simulated fault, so it cannot be regarded as a reproduction of the observation by Pacheco and Sykes[7] that natural earthquakes have a b-value of 1.5 above M_w 7.5.

When the earthquake rupture breaks the sticking points, they are called subevents. The sub-events making up a large earthquake are in many ways similar to smaller independent earthquakes. The sub-events have a perfect power-law distribution of magnitudes and are distributed uniformly and randomly over the fault plane. The same assumptions can be found in models used to explain strong ground motion spectra.[16] The sub-events have Gutenberg-Richter distribution of number, N, larger than magnitude, M,

$$\log_{10} N = A - bM.$$

Given these properties, the scaling of average slip with rupture dimensions are readily derived. First, the number of magnitude 0 sub-events is related to the maximum expected magnitude sub-event, $M_{\max}$:

$$A = bM_{\max}.$$

Magnitude M may be related to the seismic moment, M_0,[4] of each sub-event:

$$M = 2/3 \log_{10} M_0 + K.$$

The expected total moment in a rupture is found by integrating over magnitude from minus infinity up to the maximum magnitude, $M_{\max}$;

$$M_0 = \frac{2b}{3 - 2b} 10^{3/2(M_{\max} - K)}.$$

The critical property that allows average slip to be related to rupture dimensions is that the sub-events are distributed on the fault plane in a manner not influenced by dimensions of the earthquake rupture. The sub-events need not have any particular radius, and can be pictured as small patches of slip randomly distributed over the fault plane. The number of sub-events larger than magnitude 0, A, may then be related to the rupture length, L, width, W, and the one-dimensional spatial box-counting dimension d,

$$A = C(LW)^d.$$

If the sub-events are uniformly distributed over the fault plane, $d = 1$ and the constant C may be identified as a sub-event concentration, with units of number per unit area. The equations can be combined to relate total seismic moment to fault dimensions and the other parameters:

$$\log_{10} M_0 = \frac{3d}{2b} \log_{10}(LW)$$
$$+ \frac{3}{2b}(\log_{10} C - bK)$$
$$+ \log_{10}\left(\frac{2b}{3 - 2b}\right).$$

If there is no clustering of sub-events, $d = 1$, and b is also typically unity, so the 3/2 in this expression is in remarkable agreement with the 1.47 fit to data compiled by Wells and Coppersmith[14] and shown in Figure 4. Previous explanations[4,15] of this relationship based upon elasticity theory of homogeneous solids have assumed that $\overline{D} \propto W \propto L$, a relationship not supported by these data. The data set on Figure 4 set omits events from oceanic transforms and subduction zones. Rupture area is most often constrained with aftershock distributions, seismic moment with seismic moment tensor models of teleseismic waveforms. The small dots on the plot are simulated ruptures from a model with $C = 2000 km^{-2}$ and $b = d = 1$

The scatter in the relationship between rupture area and moment is substantial, about a factor of 10 in moment or 6 in rupture area. An interesting thing to note about the simulated ruptures plotted with the data on Figure 4 is that the

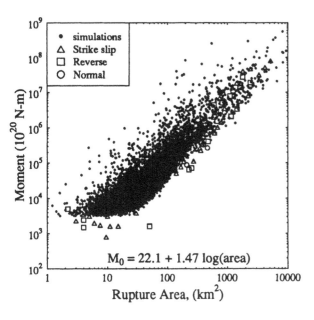

FIGURE 4 A scatter-plot of seismic moment versus rupture area for a data set compiled by Wells and Coppersmith,[14] along with their least-squares fit of the data. The small symbols are simulated earthquakes.

simulations reproduce the scatter in the relationship as well as the average trend. The scatter arises because each earthquake rupture on the plot, while including a huge number of sub-events, is dominated by the moment release of the few largest sub-events. The presence or absence of the single largest sub-event will typically change the total moment by a factor of two, and the other large sub-events make the scatter larger than that. Another important thing to note about both the data and the simulations is that the amount of scatter is the same independent of event size. This arises in the simulations because the rupture process and irregularities introduced by the sub-events is essentially scale invariant. The low magnitude simulated events are much more numerous than the high magnitude ones, so the visual spread of the points at low magnitudes is greater.

The seismic moment can be expressed in terms of shear modulus μ and average slip $\overline{D}$:

$$M_0 = \mu L W \overline{D},$$

which implies

$$\log_{10} \overline{D} \propto \left(\frac{3d}{2b} - 1 \right) \log_{10}(LW).$$

Taking $d = 1$ and a typical b-value of 1 with W constant implies $\overline{D}$ should be proportional to $\sqrt{L}$. If the sub-events are clustered with $d = (2/3)b$, the constant slip of W-scaling could result. L-scaling would occur if $d = (4/3)b$.

Figure 5 shows the same set of earthquake rupture lengths and estimated average slips as was depicted in Figure 1. The additional points on the graph are from

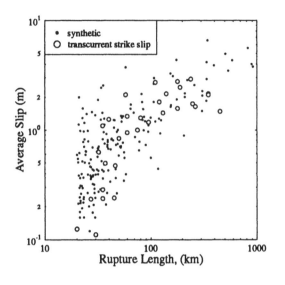

FIGURE 5 Earthquakes with estimated average slip as in Figure 1 plotted with average slip from simulated ruptures "estimated" from rupture length and seismic moment.

a simulated set of ruptures on a fault 20 km wide and very long. The concentration of sub-events C was one quarter as great as it was for the simulations shown on Figure 4, a difference that may have to do with the assumptions used to estimate areas for the earthquake data on the figure. As for Figure 1, it was assumed that all the events had a rupture width of 20 km, and the areas were computed from their lengths using $\mu = 3 \times 10^{10}$ Pa. Only events with rupture lengths greater than 20 km are plotted. The assumptions were applied to the synthetic ruptures in the same way as they were to the earthquake data.

The synthetic points on Figure 5 have an even greater scatter than the natural events, but they are most numerous on those parts of the plot occupied by natural events, and they seem to reproduce the change in slope seen in the data. In the case of the synthetics, the change in slope is enhanced by a combination of several factors. Large numbers of small magnitude events are only represented by those few having rupture lengths greater than 20 km. Many of these ruptures actually did not break the full width, so their average slips are underestimated. The synthetics do exhibit some changes in scaling when ruptures become large enough to break the full fault width. This occurs when rupture lengths are a few times the fault width. In general, the average slip exhibits such a large scatter that these effects are difficult to show with only a few events. The largest ruptures show the $\sqrt{L}$ scaling predicted analytically.

CONCLUSIONS

Highly heterogeneous slip in large earthquakes has several important consequences for the scaling of average slip with rupture dimensions. The average slip is proportional to a power of the total rupture area, with the power related to the statistical properties of the slip distribution. This proportionality is not dependent upon the fault segments being connected, and does not require that the rupture process allow the fault ends to communicate with the middle. Heterogeneity of slip could be caused by irregular material properties, geometry, or history of rupture. This model is compatible with physical arguments that faults should be pinned at the base during rupture, but it also agrees with observations that large earthquakes exhibit increasing average slip with increasing rupture length. If large earthquakes are made up of compact sub-events, and the distribution of seismic moment among the sub-events follows a Gutenberg-Richter relation, then average slip in large earthquakes should be proportional to $\sqrt{L}$.

A simple simulation of earthquake rupture was conducted which allowed strong points on the fault to act as barriers to rupture as well as asperity sub-events. Simulated ruptures reproduced the observed scatter in the relationship between rupture area and seismic moment. These simulations also showed a change in average slip for ruptures longer than the fault width, reproducing some features of natural earthquakes on a plot of average slip vs. length of rupture. The simulated earthquakes have a distribution with magnitude very similar to observed magnitude frequency distributions, and show the expected proportionality between rupture length and average slip in large earthquakes.

ACKNOWLEDGMENTS

This manuscript has benefited from my discussions with John Anderson, Roger Bilham, Frank Evans, John Rundle, Steven Jaumé, Paul Bodin, and Carl Kisslinger.

REFERENCES

1. Aki, K. "Characterization of Barriers on an Earthquake Fault." *J. Geophys. Res.* **84** (1979): 6140–6148.
2. Hartzell, S. H., and T. H. Heaton. "Inversion of Strong Ground Motion and Seismic Waveform Data for the Fault Rapture History of the 1979 Imperial Valley, California Earthquake." *Bull. Seism. Soc. Am.* **73** (1983): 1553–1583.
3. Huang, J., and D. L. Turcotte. "Fractal Distributions of Stress and Strength and Variations of b-Value." *Earth and Planet. Sci. Lett.* **91** (1988): 223–230.
4. Kanamori, H., and D. L. Anderson. "Theoretical Basis of Some Empirical Relations in Seismology." *Bull. Seism. Soc. Am.* **65** (1975): 1073–1095.
5. Kanamori, H., and G. S. Stewart. "Seismological Aspects of the Guatemala Earthquake of February 4, 1976." *J. Geophys. Res.* **83** (1978): 3427–3434.
6. Mendoza, C., and S. H. Hartzell. "Aftershock Patterns and Main Shock Faulting." *Bull. Seism. Soc. Am.* **61** (1988): 1438–1449.
7. Pacheco, J. F., and L. R. Sykes. "Seismic Moment Catalog of Large Shallow Earthquakes, 1900 to 1989." *Bull. Seism. Soc. Am.* **82** (1992): 1306–1349.
8. Romanowicz, B. "Strike-slip Earthquakes on Quasi-vertical Transcurrent Faults: Inferences for General Scaling Relations." *Geophys. Res. Lett.* **19** (1992): 481–484.
9. Romanowicz, B., and J. B. Rundle. "On Scaling Relations for Large Earthquakes." *Bull. Seism. Soc. Am.* **83** (1993): 1294–1297.
10. Scholz, C. H. "Scaling Laws for Large Earthquakes: Consequences for Physical Models." *Bull. Seism. Soc. Am.* **72** (1982): 1–14.
11. Scholz, C. H. "A Reappraisal of Large Earthquake Scaling." *Bull. Seism. Soc. Am.* **84** (1994): 215–218.
12. von Seggern, D. "A Random Stress Model for Seismicity Statistics and Earthquake Prediction." *Geophys. Res. Lett.* **7** (1980): 637–640.
13. Wald, D. J., and T. H. Heaton. "Spatial and Temporal Distribution of Slip for the 1992 Landers, California Earthquake." *Bull. Seism. Soc. Am.* **84** (1994): 668–691.
14. Wells, D. L.. and K. J. Coppersmith. "New Empirical Relationships Among Magnitude, Rupture Length, Rupture Width, Rupture Area and Surface Displacement." *Bull. Seism. Soc. Am.* **84** (1994): 974–1002.
15. Wyss, M. "Estimating Maximum Expected Magnitude of Earthquakes from Fault Dimensions." *Geology* **7** (1979): 336–340.
16. Zeng, Y., J. G. Anderson, and G. Yu. "A Composite Source Model for Computing Realistic Synthetic Strong Ground Motions." *Geophys. Res. Lett.* **21** (1994): 725–728.

John B. Rundle,† **William Klein,**‡ **and Susanna Gross***
†Department of Geology and CIRES, University of Colorado, Boulder, CO 80309
‡Polymer Center and Department of Physics, Boston University, Boston, MA 02215
*CIRES, University of Colorado, Boulder, CO 80309

Rupture Characteristics, Recurrence, and Predictability in a Slider-Block Model for Earthquakes

A variety of authors have shown that slider-block models capture some of the dynamical properties possessed by natural fault systems. These properties include: (1) a scaling (power law) region in the curve representing event frequency as a function of magnitude and area; (2) periodic, quasiperiodic, and nonperiodic behavior; and (3) space-time clustering of events. Since clusters (avalanches) of failed blocks are the model events corresponding to earthquakes on faults, the space-time development of clusters can be examined for clues to the way rupture occurs in real earthquakes. The space-time distribution of stress, elastic energy, and slip, as well as changes in these fields during cluster formation, can also be studied as a guide for understanding similar quantities in nature. A previous paper found that model calculations led to synthetic events that had source-time functions and spatial patterns of slip similar to those found in real events. However, common assumptions in the literature about constancy of rupture velocity were not supported by the simulation data. Asperitylike slip distributions in these simulations arise from strong elastic coupling, rather than the spatially heterogeneous frictional strength often inferred for real faults. Simulation results indicate that "characteristic" earthquakes can be produced as a consequence of the nonlinear dynamics, and are clearly dynamic

Reduction & Predictability of Natural Disasters, Eds. Rundle, Turcotte, & Klein,
SFI Studies in the Sciences of Complexity, Vol. XXV, Addison-Wesley, 1996

states of the lattice that are correlated in space and time. Spatially averaged "static" stress drops for synthetic events are significantly smaller than the "dynamic" stress drops associated with the block readjustment process, due to the fact that much of the stress during sliding is redistributed to neighbors, and only a portion is actually lost. As the coupling spring constant increases, the average static stress drop decreases. Segmentation on faults may therefore be a result of the nonlinear dynamics as well as being due to geometric properties of fault systems. Results from the slider-block models may not have general validity due to the extremely simple nature of the model. Finally, these slider-block systems share other important similarities to ferromagnetic, thermal, sandpile, percolation, and other systems that exhibit scaling and critical phenomena.

1. INTRODUCTION

Understanding the physics of earthquakes is complicated by the fact that the large events of greatest interest recurr at a given location along an active fault only on time scales of the order of hundreds of years (e.g., Richter,[43] Kanamori,[25] and Pacheco et al.[41]). To acquire an adequate data base of similar large earthquakes, therefore, requires the use of historical records, which are known to possess considerable uncertainty. Moreover, instrumental coverage of even relatively recent events is often inconsistent, and network coverage and detection levels can change with time.[16] Understanding the details of the rupture process is further complicated by the spatial heterogeneity of elastic properties, the vagaries of near-field instrumental coverage, and other factors (see, for example, Heaton[19] and Kanamori[26]). We are motivated to search for a method to provide insight into details of the rupture process that is complementary to the usual observational techniques (e.g., Kanamori[26]).

For many problems in statistical mechanics, numerical simulation is the only practical means of obtaining experimental information about the behavior of a nonlinear system.[34,31,72] Numerical simulation has been used extensively to study earthquakes in the recent past.[5,12,45,47,46,51,58] In the case of earthquake faults, the long time scales involved, comparable to the human lifetime, and the (at present) unpredictable nature of the events make earthquakes difficult to study systematically in the field. For this reason, it is most advantageous to develop simulation techniques so that the physics of earthquakes can be studied easily in the computer.

In developing numerical simulation techniques, there are several issues that arise. The first is the use of massless cellular automaton models, instead of models with mass, whose dynamics must be obtained by solving a set of coupled differential equations. The original Burridge-Knopoff model[11] and its CA successors[44] were not motivated by a first principles derivation, but only as an approximation to a fault. Moreover, the relative importance of inertia in the rupture process, which is responsible for generating seismic radiation, is still not clear. Kanamori and Anderson[24] estimated that the seismic efficiency η, which measures the fraction of energy in the earthquake lost to seismic radiation, is less than 5%–10%. CA models are, in fact, dynamical maps, which are known to have a fundamental connection to the dynamics generated by the associated differential equations.[3,60] Finally, it was shown by Nakanishi[35] that massless CAs can be constructed that give the same quantitative results as the massive slider-block models examined by Carlson and Langer.[12] These results include not only the rupture patterns, but also the scaling statistics. There is also clearly a time evolution process in the CA, with a separation of loader plate and source time scales, just as in models with inertia. The physical interpretation of the CA time scales is discussed by Gabrielov et al.[14] To summarize, this issue of the presence or absence of inertia has been examined in the literature and shown to have little significance.

In every simulation, massive or massless, there is a series of fundamental approximations. For example, faults are usually assumed to be planar; boundary conditions might be periodic, or fixed at the ends; interactions between fault segments might be nearest neighbor, $1/r^3$, or long range of some other variety. Friction might be simple stick slip (Mohr-Coulomb); might have simple but unrealistic velocity dependence as in the original Burridge-Knopoff model; might be based on overly idealized laboratory friction experiments on clean, dry, dust-free sliding surfaces; might ignore the effects of pore water; or might ignore effects associated with variable normal stress on the fault. All of these various models have their advantages and disadvantages, their adherents and opponents.

In this paper we use the simplest possible model, which we term a "toy" model. There are several important reasons for studying such a toy model. Any dynamical model, which includes both differential equations as well as maps, gives rise to an observed dynamical behavior as a consequence of the physical nature of the model. It is of fundamental interest to ask what minimal set of physics must the model possess in order to demonstrate dynamical effects that are seen in nature. We are therefore making a fundamental distinction between models that previous experience tells us are "realistic," and "minimalist" models which nevertheless demonstrate important similarities to natural phenomena. The toy model that we examine in this paper is of the latter category. One expects "realistic" models to match natural fault dynamics well.

The true value of the toy model lies exactly in the somewhat surprising fact that a significant body of natural phenomena are reproduced by the dynamics, including both scaling of earthquake distributions and, as demonstrated here, "characteristic earthquakes." This fact leads to the important conclusion that the observed phenomena may either have several origins, or that the true origin is not in the physics included in the more "realistic" model. In short, the objective of this paper is to ask the question, how little physics can one include in a model and still get interesting, unexpected results? For example, massless CA slider-block models obtain the same kinds of scaling distributions for earthquakes as do massive models which possess inertia. Moreover, neither massive nor massless models include elastic waves, attenuation, viscoelasticity, etc. The logical question is then to what extent the presence of inertia, waves, and so forth play any role in the scaling properties of the phenomena. Another example, which is a major subject of this paper, is that massless slider-block models with a uniform failure (strength) threshold can reproduce much of the phenomena associated with earthquakes,[29] including segmentation phenomena, "characteristic earthquakes," and space-time clustering of events. Natural characteristic earthquakes are usually thought to be associated with a spatially heterogeneous failure threshold or the geometry of a fault, but the results obtained here raise questions about the true physical origin of these phenomena.

The toy model we use here has been introduced elsewhere.[37,40] Other papers have focused on examining the scaling properties of the model and establishing its universality class.[37,54] Here we explore the phenomenology of the model, beginning with a short description of the model, then give a graphical display of a series of representative results. We compare these to observed data from seismology. In general, one obtains Heaton-type pulses[19] rather than cracklike solutions,[28] because healing takes place immediately after a block slides, rather than at a delayed time after slip is complete everywhere.

2. MODEL

Our model[44] consists of a massless cellular automaton (CA) version of the model originally proposed by Burridge and Knopoff[11] (see Figure 1). Later versions of have been proposed by, among others, Nakanishi,[36] Brown et al.,[10] Narkounskaia et al.,[37] and Olami et al.[40] A network of massless blocks sliding on a frictional surface are connected to nearest-neighbor blocks by coupling springs with spring constant K_C. A loader spring with spring constant K_L connects each block to a loader plate moving with a velocity V, increasing the force on each block at a steady rate. In the original BK model with massive blocks, the frictional force has a velocity dependence that produces lower stress with increased sliding velocity. Once a static force threshold σ^F is equaled or exceeded, the full inertial equations of motion are solved in the BK model to obtain the slip of each block.

The solution of coupled differential equations makes the simulation of systems with large numbers of elements extremely difficult. Moreover, it is important to simulate as large a system as possible to eliminate the effects of the finite size of the lattice. By contrast, the CA approach for these systems can be made far more computationally efficient since only simple algorithms are used. In slider-block models, eliminating the mass in turn eliminates the possibility of waves and radiation damping. If the seismic efficiency is low, of the order of 5%–10% as is

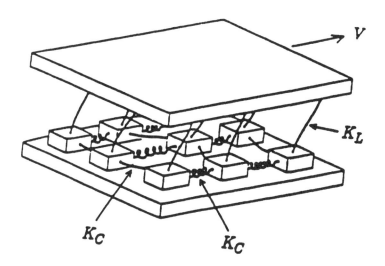

FIGURE 1   Picture of the two-dimensional slider-block model.

generally assumed (e.g., Kanamori and Anderson[24]), a case can be made that inertia is less important than other entropic and internal energy effects.

Only the simplest kind of stick-slip friction is used in our CA model. "Clusters" or avalanches of failed blocks can appear because an initially unstable block increases the force on its neighbor blocks as a result of the coupling springs, thereby inducing instability in its neighbors. Each of the clusters of failed blocks represents an earthquake in the simulation. The Hamiltonian for the system of blocks and springs is:

$$H = \frac{1}{2}\sum_i \{K_L(s_i - Vi)^2 + \frac{1}{2}K_C\sum_j [s_j - s_i]^2\}. \tag{1}$$

In Eq. (1), s_i represents the total slip of block i, V is the loader plate velocity, t is time, and $\sum_j$ is carried out over the blocks that are nearest neighbors to block i. The sum $\sum_j$ excludes site i, and the factor $(1/2)$ corrects for the double counting of site i. Equation (1) can be written as:

$$\begin{aligned}
H &= \frac{1}{2}\sum_i \{K_L(s_i - Vt)^2 + K_C\sum_j [(s_j - Vt) - (s_i - Vt)]^2\} \\
&= \frac{1}{2}\sum_i \{K_L\phi_i^2 + K_C\sum_j [\phi_j - \phi_i]^2\}
\end{aligned} \tag{2}$$

where $\phi_i = s_i - Vt$ is the negative of the slip deficit of block i.

The force σ_i (or stress if unit area is assumed) on block i is :

$$\sigma_i = -\frac{\partial H}{\partial \phi_i} = -\{K_L \phi_i + K_C \sum_j [\phi_j - \phi_i]\}. \tag{3}$$

A rule that generates evolution through time must be specified, this being the friction law. For the simulations here, we adopt the Mohr-Coulomb law with a spatially constant static failure threshold σ^F, and a spatially constant residual stress σ^R. Upon slip, when $\sigma_i \geq \sigma^F$, each block jumps forward a distance $\Delta s_i = (\sigma_i - \sigma^R)/K_T$, $K_T = K_L + 4K_C$, whereupon it sticks (heals). Different Monte Carlo prescriptions have been used in other realizations of the model.[54] The statistical distributions obtained as well as qualitative properties of the results are independent of the time evolution rule used, suggesting that there may be some universal aspects of the simulations.

The Hamiltonian (1) for a slider-block model can be formally obtained from the expression for the energy change due to slip on a planar fault surface in an infinite elastic medium. The expression for the energy contained within an elastic medium due to quasi-static slip on the fault surface S is:

$$H_{EL} = -\int_S d^2\mathbf{x} \left\{ p(\mathbf{x})s(\mathbf{x},t) + \frac{1}{2}\int_S d^2\mathbf{x}' T(\mathbf{x} - \mathbf{x}')s(\mathbf{x},t))s(\mathbf{x}',t)) \right\}. \tag{4}$$

The important assumptions are: (1) the constant applied stress field $p(\mathbf{x}) = 0$; (2) a shift to a moving reference system, so that slip s is replaced by slip deficit ϕ; and (3) the Green's function $T(\mathbf{x} - \mathbf{x}')$ falls off sufficiently rapidly as $|\mathbf{r}| = |\mathbf{x} - \mathbf{x}'|$ increases. This last assumption implies that the system (earth) is not ideally elastic, and that there is some infrared "cutoff." In fact, the earth is not ideally elastic, inasmuch as wave attenuation, material creep, plasticity, fluid flow, and other processes are important in fault zones and throughout lithosphere. The issue of whether there exists an infrared cutoff is equivalent to the problem of whether the elastic interaction is screened by competing interactions due to defects in the solid.[31] An example of a Green's function with an infrared cutoff is the asymptotic form $e^{-\alpha r}/r^3$, rather than $1/r^3$, the form appropriate to a perfectly elastic solid. The effect a nonzero screening parameter α has on the value of K_C/K_L is discussed elsewhere.[52]

Using a gradient expansion[30] of $\phi(\mathbf{x}',t')$ around $\mathbf{x}$, retaining terms up to the second derivative in $\mathbf{x}$, and integrating by parts, the local interaction energy H_{EL}^{loc} is obtained:

$$H_{EL} \approx H_{EL}^{loc} \equiv \frac{1}{2}\int_S d^2\mathbf{x} \left\{ k_L \phi^2(\mathbf{x},t) + k_C(\nabla\phi(\mathbf{x},t))^2 \right\} \tag{5}$$

where the spring constant densities k_L and k_C are related to the Green's function by:

$$k_L \equiv - \int_S T(\mathbf{r}) d^2 \mathbf{r} \qquad (\text{$-$0th moment of } T); \tag{6}$$

$$k_C \equiv \frac{1}{2} \int_S T(\mathbf{r}) \mathbf{r}^2 d^2 \mathbf{r} \qquad (\text{2nd moment of } T). \tag{7}$$

To obtain the slider-block energy, Eq. (5) is spatially coarse grained by averaging the integrand over squares of side length a. Then $\phi(\mathbf{x}) \to \phi_i$, and noting that $\nabla \phi(\mathbf{x}) \approx (\phi_i - \phi_j)/a$, the slider-block energy (2) is obtained. In energy functionals like Eq. (5), k_C is the "range of interaction," and k_L is the "mass" (e.g., Ma,[30,31] Amit,[2] and Yeomans[72]).

Finally, it should be pointed out that some authors prefer models with considerably more physical detail, specifically with infinite range interactions and a five-parameter friction law (e.g., Ben-Zion and Rice[7]). While that approach has advantages, it has the disadvantages that (1) infinite range interactions mean always having to deal with boundary conditions and finite size effects, and (2) it is often difficult to identify the physical origin of various phenomena. Our approach is to begin with as simple a model as possible (minimal model) and progressively add more levels of detail as the effects of the different parameters are clarified.

Since a major concern at present is computational and analytic simplicity, we use a slider-block CA model with spatially constant failure threshold σ^F and residual stress σ^R to simulate seismicity on a fault. The effect of randomness in the model is examined elsewhere.[56] The time evolution of the CA model is generated by a jump rule giving the position of the block as a function of the state of stress on the block. The jump rule for the ith block is given by:

$$s_i(t+1) = s_i(t) + J(\sigma_i)\Theta(\sigma_i - \sigma^F) \tag{8}$$

where $\Theta(x)$ is a Heaviside step. The jump function $J(\sigma_i)$ is

$$J(\sigma_i) = \frac{\sigma_i - \sigma^R}{K_L + 4K_C}. \tag{9}$$

In the work described in this paper, the simulations were started with the blocks having random initial positions, on lattices of 100×100 blocks. The calculations proceed as follows:

1. The loader plate was moved (updated) until the stress on the least stable block reached threshold ($\sigma_i = \sigma^F$).
2. The position of the least stable block is adjusted using Eqs. (8)–(9) (synchronous updating).
3. A "Monte Carlo" sweep (iteration) through the entire network (lattice) is carried out and the stress on each block is calculated using Eq. (3).

4. The position of all blocks at or above threshold ($\sigma_i \geq \sigma^F$) are simultaneously adjusted according to Eqs. (8)–(9).
5. Further sweeps and block position adjustments are carried out until all of the blocks are in states with $\sigma_i \leq \sigma^F$.
6. The loader plate position is again updated according to step 1.

The time evolution process described in steps 1–6 is contained in Eqs. (3), (8), and (9). This model is a $V \to 0$ model, because the loader plate moves so slowly that there can only be one initiator block in each avalanche cluster of failed sites. This is the same basic model described by Narkounskaia et al.[37] and Olami et al.[40] Growing avalanche clusters will never coalesce into larger clusters, so that each cluster in this (nearest-neighbor) model grows outward from only one site. This is evidently an important distinction among various slider-block models with respect to scaling exponents and limit-cycle behavior.

---

3. SIMULATION RESULTS

An examination of simulations using this model in an earlier paper[52,53] demonstrated that (1) the model generates synthetic seismicity with space-time clustering, with magnitude-time, moment-time, and magnitude-frequency frequency distributions similar to natural faults and fault systems; (2) rupture characteristics depend strongly on the choice of failure rules and model parameters; and (3) asperitylike features arise in models even with spatially uniform threshold strength, and are a result of strong elastic coupling ($R^2 = K_C/K_L$ large). In this paper, we are concerned more with recurrence patterns, rupture characteristics, and implications for predictability.

Results of our simulations are shown in Figures 2–17. In all these calculations, a spatially constant threshold $\sigma^F = 350$ was used, together with a spatially constant residual stress $\sigma^R = 10$. Values used for spring constants were $K_C = 1$ (always), and $K_L = 1$ or $K_L = .04$ (as noted in the figure captions and described below). One objective of the simulations was to examine the extent to which changes in only one parameter, the coupling ratio $R^2 = (K_C/K_L)^{1/2}$, can determine the fundamental phenomenology of the dynamical process. A second major objective was to evaluate the extent to which large (synthetic) earthquakes are predictable, in the sense that major characteristics are preserved from event to event. In all cases, simulations were begun from random initial conditions, then the models were run for at least several hundred thousand events to eliminate transients before data were obtained.

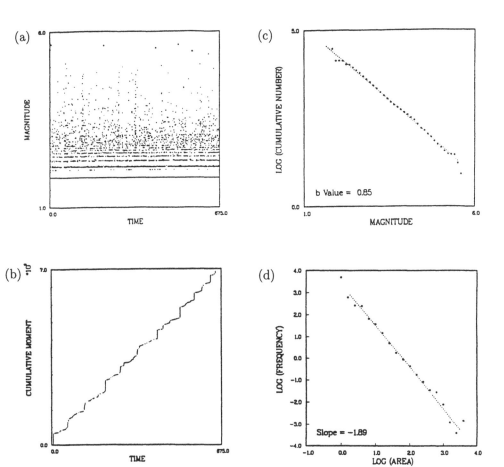

FIGURE 2   Results from simulation with $R^2 = K_C/K_L = 1$, $K_L = 1$, on a 100 × 100 lattice of blocks. (a) Magnitude-time plot of synthetic events, with magnitude as defined in the text. (b) Cumulative moment-time plot. (c) Cumulative magnitude-frequency relation for all events. Note that b-value of .85 should not be compared with Gutenberg-Richter values for real data since magnitude is defined as $\log_{10}$ (moment). (d) Probability density for frequency of occurrence plotted against area of event. Note that slope of -1.89 is close to the slope of -2 for real data, and is independent of assumptions about constants in the magnitude-frequency relation.

In Figure 2 we show simulated earthquake statistics for 20,000 events in a model with $R^2 = 1$. In these and subsequent figures, the seismic moment M_o for an avalanche of failed blocks is defined as:

$$M_o = K_L \sum_{\text{(all failed blocks)}} \{\text{slip}\} = K_L \text{ (average slip) (total area)} \qquad (10)$$

where the summation is carried out over all failed blocks in the avalanche. Magnitude M is defined by:

$$M = \log_{10}\{M_o\}. \qquad (11)$$

This expression differs by constant factors from the moment magnitude relation for natural events. For the moment, these details are not of interest, we are interested only in preserving the logarithmic dependence of magnitude M on moment M_o.

The top left corner plot of Figure 2 shows M as a function of time for a model time interval of 675, where time is defined as the position of the loader plate. The apparent "horizontal solid lines" in the lower portion of the plot are actually individual dots representing, from the bottom, the one-block events, next the two-block events, and so forth, until at higher magnitudes where events with many failed sites are resolved into single isolated points. That the one-site events appear as a horizontal line is due to limitations in resolution of the laser printer. The first three largest magnitude events, near the top of the plot, are approximately equidistantly spaced in time, and are followed by a sequence of ~3 large events clustered more closely in time. Space-time clustering of large events is an important property observed in natural seismicity[23] (see also data below). The large simulation events will be examined in more detail below.

Next in Figure 2, the left lower plot (b) shows cumulative magnitude as a function of time, and closely resembles many such plots of real data (e.g., McNally[32]). The top right plot (c) is the cumulative number N of events as a function of magnitude (Gutenberg-Richter plot) for the events, and it can be seen that these simulation data conform to the relation:

$$\log_{10} N = 5.1 - .85M. \qquad (12)$$

The b-value of .85 is actually not in agreement with the value near 1 observed in nature, because the observational data uses a relation between M and $\log_{10} M_o$ different by a factor of $2/3$. However, it may be significant that the relation between $\log_{10}\{\text{Frequency}\}$ and $\log_{10}\{\text{Area}\}$ has a slope $(-1.89$: lower right plot (b)). Note that the "frequency" in this frequency-area plot is the probability density function for frequency, not the cumulative frequency. An observational-empirical value for this exponent of -2 is implied by the observational moment-magnitude scale. For the cumulative frequency, an observational value of -1 is found.

In Figure 3 we show the largest events in the form of two "recurrence plots" of event time in terms of event sequence number, similar to figures commonly used to present Parkfield data (see Bakun and McEvilly[6]; see also Scholz,[63] p. 246). Top plot shows all events larger than $M = 5.0$ for a model time interval of 2100, bottom plot shows only events larger than $M = 5.5$ for the same time interval. Temporal clustering of the large events similar to natural seismicity[23,68] can readily be seen from this plot, implying that earthquake straight-line forecasting/prediction, using time-predictable or slip-predictable ideas, is possible for only limited time intervals in this model. Moreover, the difference between top and bottom plots clearly implies that the degree to which events are "predictable" depends strongly on the events selected for plotting. The lack of predictability in these plots has also been emphasized by Savage.[61,62]

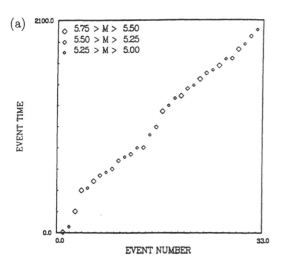

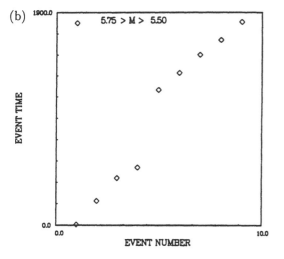

FIGURE 3   (a) Top is event time plotted against sequence number for the events with $5.75 > M > 5.0$ shown in Figure 2. (b) Same kind of plot as for (a) but only for the largest events, $5.75 > M > 5.5$. Temporal clustering is evident in both plots.

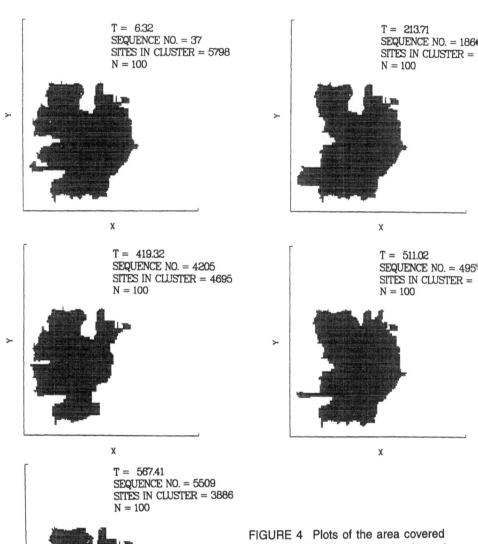

T = 6.32
SEQUENCE NO. = 37
SITES IN CLUSTER = 5798
N = 100

T = 213.71
SEQUENCE NO. = 186
SITES IN CLUSTER =
N = 100

T = 419.32
SEQUENCE NO. = 4205
SITES IN CLUSTER = 4695
N = 100

T = 511.02
SEQUENCE NO. = 495
SITES IN CLUSTER =
N = 100

T = 567.41
SEQUENCE NO. = 5509
SITES IN CLUSTER = 3886
N = 100

FIGURE 4 Plots of the area covered by the first five largest events shown in Figure 2 (a). Sequence of events proceeds from top left to top right to middle left, etc. The similarity of one event to the next can be seen, lending credence to the characteristic earthquake idea; however, event configuration is changing also. This event occurred roughly in the middle of the 100 × 100 lattice of blocks.

In Figure 4 we display the spatial configuration of the failed blocks for the first five largest events, as indicated along the top of Figure 2 (top left). These events closely resemble one another. In fact, it can be shown that slider-block models are very closely related to integrate-and-fire neural network models[21,55] Self-organization and memory are intrinsic properties of such models, and originate both from the existence of Lyapunov functions, as well as from the long-wavelength correlation of fluctuations. Patterns will emerge, persist for a period of time, and then disappear to be replaced by new patterns. The events seen in Figure 4 are reminiscent of characteristic earthquakes,[59] in which similar events repeat at regular intervals for a period of time, as exemplified by the Parkfield sequence. Moreover, it has been shown[14] that for fixed stress drop during block slip ($\Delta\sigma_i = \sigma^F - \sigma^R = \text{constant}, \sigma_i > \sigma^F$), slider-block models display only periodic behavior. The present model, which departs somewhat from these conditions ($\Delta\sigma_i = \sigma_i - \sigma^R, \sigma_i > \sigma^F$), can therefore be expected to demonstrate approximately periodic behavior over limited time spans. This is particularly true when the coupling is weak ($R^2 \to 0$) so that σ_i never rises much above σ^F and $\Delta\sigma_i = \sigma_i - \sigma^R \approx \sigma^F - \sigma^R$.

Figure 5 illustrates the analogue of the source-time moment rate function for the first three of these large events. Each event develops as an avalanche of failed blocks that occur during a series of iteration sweeps through the lattice. The moment rate as a function of iteration time is obtained by summing the slip produced by all the blocks that fail during each iteration sweep. This function is the analogue to the source time function for natural events. Figure 5 demonstrates that both the spatial pattern of block failures and the moment rate function change progressively with succeeding events. The area under the moment rate curve associated with the small "pre-shock" that is apparent at the beginning of the first event at top, grows in importance until by the third event (bottom) it represents a significant fraction of the total area under the curve associated with the "main event." Note the large fluctuations in moment rate compared to the time average in all three figures, to be contrasted with Figure 13 below.

In Figure 6 we show three-dimensional projections of the slip (a), stress drop (b), and energy decrease (c) for the first large event in the sequence, the same event as shown previously in Figures 2–5. The distinguishing characteristic of all three plots is the spatially constant nature of the changes in slip, stress drop, and energy. The spatially uniform region bounded by sharp edges in these quantities is characteristic of small values for R^2, the parameter whose inverse measures the size of spatial fluctuations in slip difference between neighboring sites (e.g., Eqs. 1–5). Conversely, as R^2 increases, spatial fluctuations decrease and the spatial distributions are smoother (e.g., Figure 14 below); in this case, slip, stress drop and energy are no longer sharply concentrated at boundaries.

Figure 7 is complementary to Figure 6, (c), illustrating the state of stress throughout the lattice before and after the first large event. The lattice average of the pre-event stress $\langle\sigma_b\rangle = 220$, the post-event average $\langle\sigma_a\rangle = 176$. Since the event

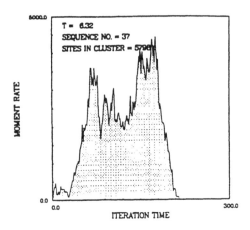

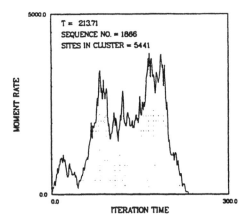

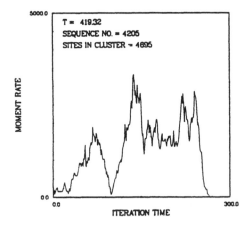

FIGURE 5   Moment rate plotted against iteration (event) time for the first three large events shown in Figures 2. Evolution of the spatial configuration of the events is associated with evolution of the moment rate function.

3 – D PLOT OF SLIP

(a)
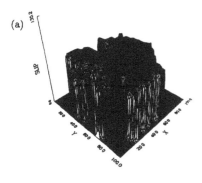

STRESS DROP

(b)
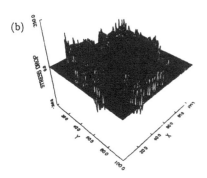

ENERGY DECREASE

(c)
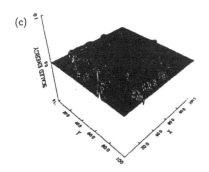

FIGURE 6   (a) Three-dimensional perspective plot of the slip in the first large event shown in Figure 4. (b) Plot of stress drop. (c) Energy decrease. All quantities are nearly constant as a function of spatial position.

(a)

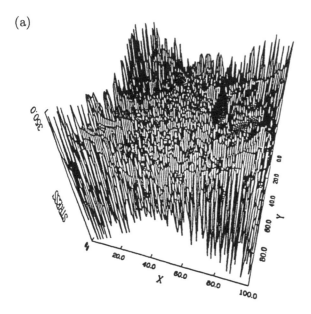

(b)

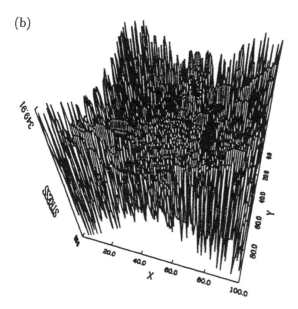

FIGURE 7 Three-dimensional perspective plots of spatial dependence of (a) pre-event stress field and (b) post-event stress field associated with the time just before and just after the first large event shown in Figure 4. There is little apparent spatial smoothing of the stress field as a result of the event.

has area 5798 sites, the average static stress drop $\Delta\sigma = 75.9$. Comparison with Figure 15 having $R^2 = 25$ and $\Delta\sigma = 12.6$ indicates that larger values of R^2 are associated with an increased fraction of stress redistribution relative to stress lost, and therefore smaller values of $\Delta\sigma$ (see below). Also, it is important to notice that both lattice-average pre-event $\langle \sigma_b \rangle = 220$ and post-event $\langle \sigma_a \rangle = 176$ stresses are substantially less than the threshold $\sigma^F = 350$. In these simulations, the lattice-average stresses fluctuate in time about a value that is substantially less than σ^F. The simple view of stress on faults rising uniformly to the threshold, followed by a decrease of stress to zero, is not correct for these simulations.

Details of the process by which simple periodicity of the largest events is disrupted and simple predictability eventually destroyed is shown in the time development of rupture patterns in Figures 8–10 for the first three large events. In these figures, the lightly shaded region is all of the sites that eventually fail; the darker region represents the cumulative sites that have failed by the iteration sweep shown in the upper left of each panel; and the darkest sites represent the sites actively failing on the iteration sweep indicated. These darkest sites represent the "rupture front." The plot on the upper right of the figures is an early iteration-time snapshot of the rupture front after 60-iteration time intervals. Rupture progression is documented at further iteration times of 120 and 180. It can be seen that the rupture process in this model is consistent with the slip pulse model of Heaton,[19] in that a narrow region at the rupture front sweeps rapidly over sites on the fault, leaving slipped sites behind the rupture front locked again. Somewhat unlike the simple Heaton model, however, the rupture front is not necessarily a continuous line of slipping sites, but is irregular, sometimes consisting of only a very few sites. After about 200 iteration steps, slip on the model fault is complete. The location of the initial rupture site changes somewhat from event to event, and details of rupture front evolution change progressively in the three events. However, there are clearly common features: the initial site remains in the lower left corner, rupture progresses from lower left to upper right, and occurs at a well-defined rupture front.

Figures 11–17 are a series of plots similar to Figures 2–10, but for a model in which $R^2 = 25$ ($K_C = 1, K_L = .04$). In Figure 11 (a), there is less evidence of temporal clustering than was seen in Figure 2 (b). However, there are still a number of large events similar to those seen for the previous case, and these events are the subject of the succeeding figures. Among obvious differences between Figures 11 and 2, the magnitude-frequency plot is not well represented by a relation of the form (12), unlike the result in Figure 2; and the frequency-area relation is not as good a power law as that shown in Figure 2, and has a lower average slope (~ -1.6). In fact, the magnitude-frequency relation at top right of Figure 7 is somewhat similar to that observed for isolated segments of major fault zones such as the Prince William sound section of the Alaska subduction zone (e.g., Scholz,[63] p. 188). This type of magnitude-frequency relation has been used to justify the characteristic earthquake model of earthquake recurrence along a fault.

In Figure 12 we show Parkfield-type recurrence plots for the largest events over a time span of 6500, similar to Figure 3. The upper panel shows all events larger than $M = 4$, the lower panel shows all events larger than $M = 4.25$. Temporal clustering of the largest events is more pronounced than for smaller events. Similar

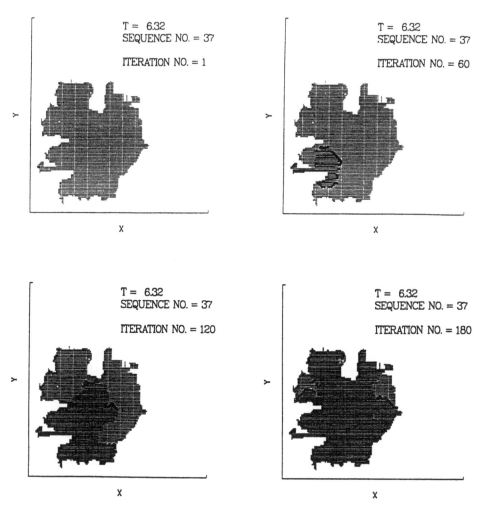

FIGURE 8   Space-time pattern of failure (rupture) progression for the first large event shown in Figures 2 and 4. Initiation point (hypocenter) is shown at top left, other snapshots of failure are shown at intervals of 60 iteration time steps proceeding from upper left to bottom right. Darkest shading are sites that are failing on the indicated iteration time step, lightest shading indicates sites that have not as yet failed by the indicated iteration step, and medium dark shading indicates sites that have failed. The darkest shaded sites constitute the rupture front.

conclusions to Figure 3 concerning lack of predictability based on straight line forecasting apply.

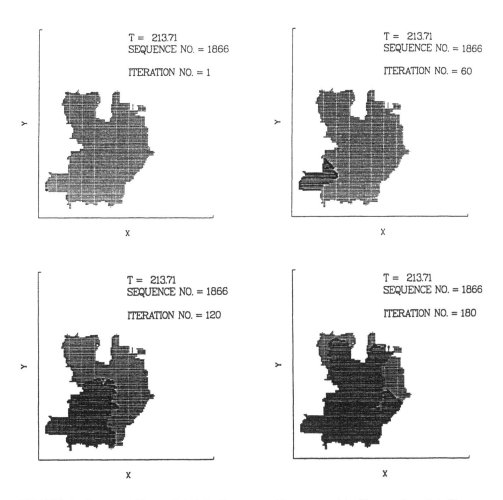

FIGURE 9   Same as Figure 8 but for the second large event in Figures 2 and 4. The similarities and differences in the rupture processes between the event shown here and in the previous figure at the given iteration time steps should be noticed.

Figure 13 is analogous to Figure 5 but for the first two large events in the magnitude-time plot in Figure 11. The source-time (moment rate) functions in Figure 13 are again calculated by summing the slip of each sliding block at a given iteration time. Unlike Figure 4, however, the moment rate functions shown here are temporally much smoother with smaller fluctuations relative to the time average. As described in association with Figure 13, the relative smoothness is due to the larger value of R^2.

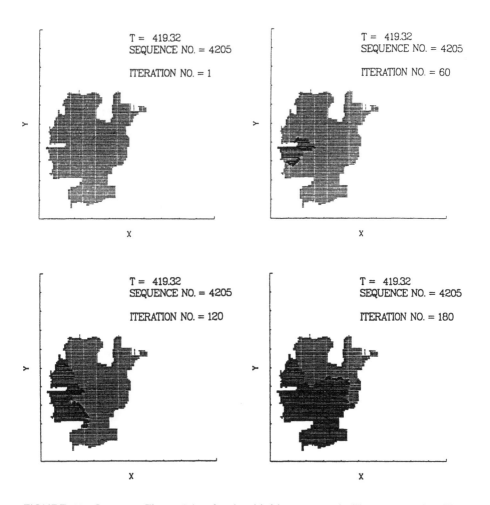

FIGURE 10   Same as Figure 8 but for the third large event in Figures 2 and 4. The similarities and differences in the rupture processes between the event shown here and in the previous figures at the given iteration time steps should be noticed.

In Figure 14 we show three-dimensional projection views of slip, stress drop, and energy decrease for the first large event shown in Figure 13. In contrast to Figure 6, in which these quantities are essentially constant over the area of failure, Figure 14 indicates substantial spatial variation of slip, stress drop, and energy change, a result of the repeated and irregular pattern of failure for individual sites. Slip and energy change are clearly smoother spatially, a result of the larger value of R^2, but stress drop is spatially heterogeneous, suggesting that the product $\Delta s_{ij}^2 R^2$

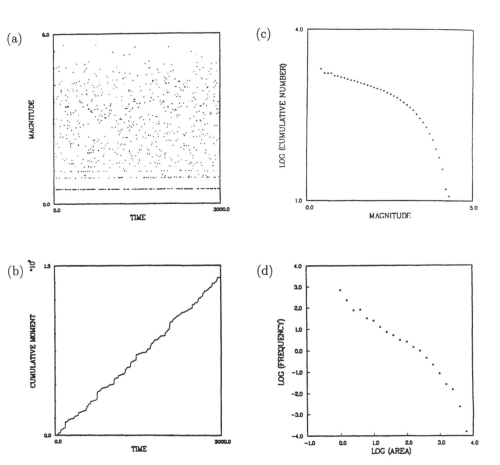

FIGURE 11   Results from simulation with $R^2 = 25$, $K_L = 1$. (a) Magnitude-time plot of synthetic events, with magnitude as defined in the text. (b) Cumulative moment-time plot. (c) Cumulative magnitude-frequency relation for all events. Unlike Figure 2, the magnitude-log frequency relation is not well represented by a straight line. (d) Probability density for frequency of occurrence plotted against area of event. Again, these synthetic data are not well represented by a line (i.e., power law for frequency-area).

of squared slip difference $\Delta s_{ij}^2 = (s_i - s_j)^2$ and R^2 varies relatively slowly with position.

Figure 15 is analogous to Figure 7. For this value of $R^2 = 25$, however, the lattice average of pre-event stress is 246, the post-event stress is 241, and the area

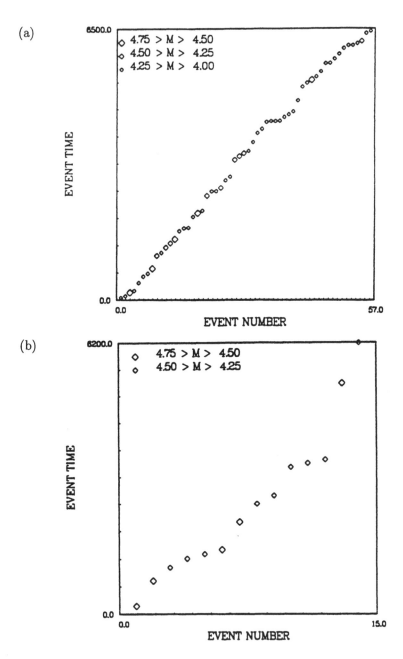

FIGURE 12 (a) Top is event time plotted against sequence number for the events with $4.75 > M > 4.0$ shown in Figure 2. (b) Same kind of plot as for (a) but only for the larger events, $4.75 > M > 4.25$. Temporal clustering is evident in both plots.

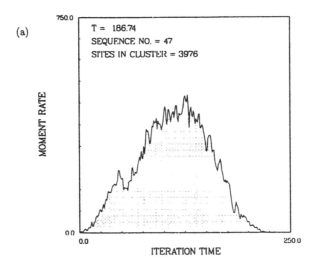

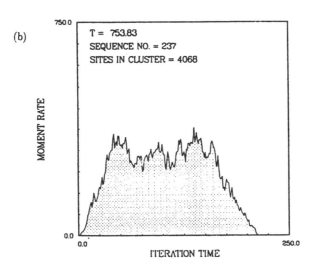

FIGURE 13 Moment rate plotted agains iteration (event) time (on left) for the first two large events shown in Figures 11 (a) for the model with $R^2 = 25$. Evolution of the spatial configuration of events is associated with evolution of the moment rate function.

3 – D PLOT OF SLIP

(a)

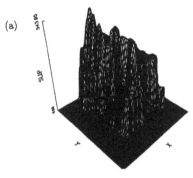

STRESS DROP

(b)

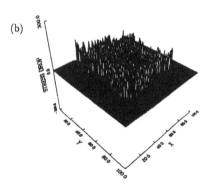

ENERGY DECREASE

(c)

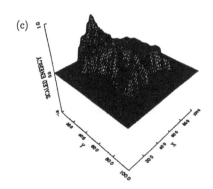

FIGURE 14   (a) Three-c
perspective plot of the s
event shown in Figure 5
drop. (c) Energy decrea
quantities vary consider
position.

(a)

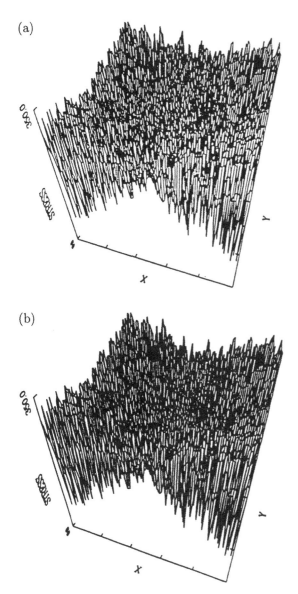

(b)

FIGURE 15 Three-dimensional perspective plots of (a) pre-event stress and (b) post-event stress for the first large event shown in Figure 13.

covered by the cluster is 3976 sites. Average stress drop $\Delta\sigma = 12.6$, much smaller in magnitude than the value $\Delta\sigma = 75.9$ found in Figure 7 for the event with $R^2 = 1$. Far more of the stress is redistributed than is lost as R^2 increases. We conclude that the ratio $\Delta\sigma/\sigma^F$ is a rapidly decreasing function of R^2. In nature, $\Delta\sigma$ is the static stress drop measured in earthquakes, and has a value of ~ 1 to 10 MPa,

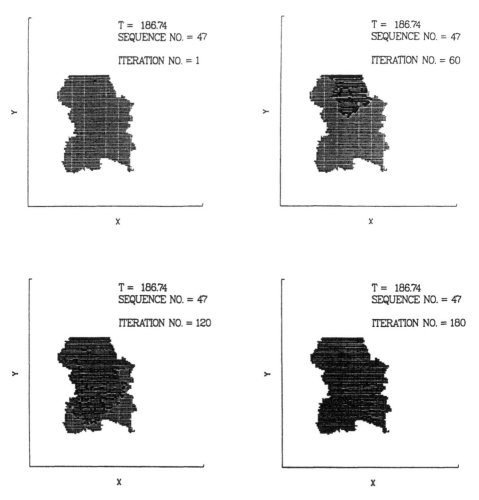

FIGURE 16 Space-time pattern of rupture progression for the first large event that was shown in Figure 13. Initiation point (hypocenter) is shown at top left, other snapshots of failure are shown at intervals of 60 iteration time steps proceeding from upper left to bottom right. Lightest shading indicates sites that are yet to fail, darker shading indicates sites that have failed, and darkest shading are sites that are failing on the indicated iteration time step. Only a few of the latter sites can be considered to constitute the rupture front, since in this model, sites fail multiple times and a considerable amount of slip continues to occur on previously ruptured sites.

while σ^F can be measured by the breaking strength of rocks, ~ 1 GPa. This small ratio $\Delta\sigma/\sigma^F$ is therefore consistent with the large value of R^2 expected for an approximately elastic earth. Again, both lattice-average pre-event $\langle\sigma_b\rangle = 246$ and post event $\langle\sigma_a\rangle = 241$ stresses are substantially less than the threshold $\sigma^F = 350$.

FIGURE 17   Same as Figure 16 but for the second large event shown in Figure 13.

Finally, Figures 16–17 are plots of time development of rupture for the first two large events shown in Figure 13, similar to Figures 8–10. The most obvious difference between the examples with $R^2 = 1$ (Figures 8–10) and $R^2 = 25$ (Figures 16–17) is that failure takes place only at the rupture front in the former, whereas failure occurs over a broad zone of failing sites behind the initial rupture front in the latter. This broad zone is a result of the relatively large value $R^2 = 25$, because most stress is redistributed to neighbors rather than lost when a block slips. Sites can fail repeatedly for large R^2, and the number of times a block can fail grows with R^2.

4. NATURAL SEISMICITY DATA

Whether our conclusions about predictability apply to earthquakes on natural fault systems, seismicity depends on the extent to which the slider-block model captures the physics of natural seismicity data. We now present evidence that the slider-block models do demonstrate some of the observed phenomena, thus implying that the simulations may incorporate part of the essential physics. In particular, we examine aftershocks of the June 28, 1992 Landers earthquake (Figure 18), and compare the probability density for temporal clustering of these events with that for temporal clustering of the simulation events.

We use aftershock data to look for evidence of clustering because instrumental coverage is uniform in space and time, and because of the large number of these events available for use. However, there is a problem in using these data for which we must make an adjustment. This problem is the inverse-time decay in aftershock numbers, the Omori effect. We therefore wish to refer the times of the real events to the times expected from the Omori relation, and using these time differences, examine the "anomalous" clustering, the clustering that is over and above that expected from the Omori effect.

In Figure 19 we show distributions of interevent times for natural, theoretical (random Poisson), and synthetic seismicity. The synthetic seismicity is from the simulation data with $R^2 = 1$ shown in Figure 2. The natural activity are Landers aftershocks (Figure 18), to which the following procedure is applied to construct the interevent distributions in Figure 19:

1. A modified Omori relation is used to fit the aftershock decay process.[15]
2. Using this modified Omori relation, the expected number of events at the times of all the aftershocks in the set are predicted. These predicted numbers of events define a measure of event time called frequency linearized time (FLT).
3. The FLT event times are used to calculate a distribution of differences in FLT between the data and the best fitting Omori relation, thus reflecting clustering independent of the overall aftershock decay.
4. The distributions of natural and simulation seismicity are divided by an exponential with the same average interevent time. The exponential distribution is the distribution that interevent times drawn from a Poisson point process must possess. This step allows us to compare the distribution of interevent times to those expected from a purely random process.
5. For ease of comparison, all distributions are normalized to an average interevent time = 1.

In other words, the net effect of this procedure is to construct a distribution of interevent times that have the expected temporally decaying Omori clustering removed, and which are normalized to a distribution expected for a random process. In this way we can look for evidence of anomalous clustering, with the more dominant clustering produced by the Omori decay removed.

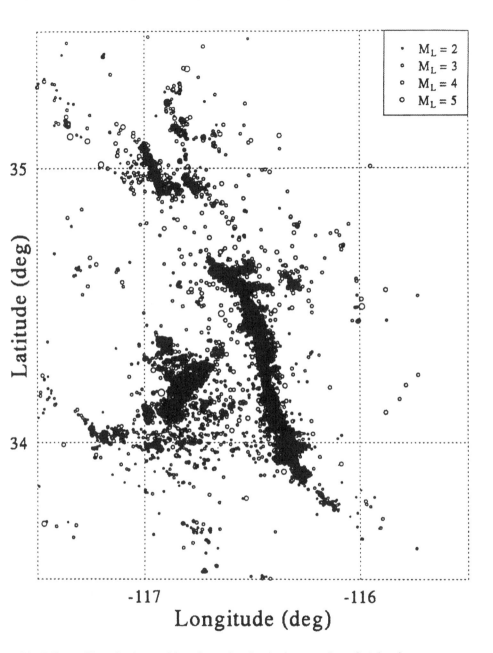

FIGURE 18   Plot of relocated Landers aftershocks in map view. Catalog from Hauksson et al.[18]

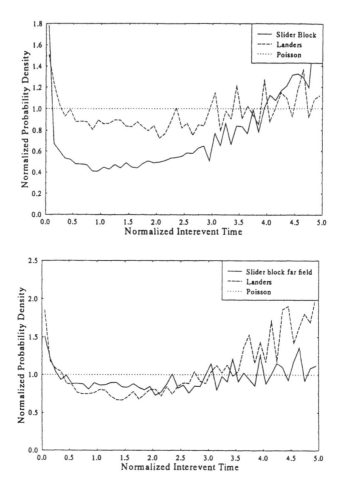

FIGURE 19   Top: Cumulative interevent times for Landers aftershocks (dashed) and simulated seismicity (solid), normalized to an exponential distribution (Poisson point process: dotted line). Since the distribution of events drawn from the Poisson point process is normalized to an exponential distribution (i.e., to itself), its distribution appears as a horizontal line at the value 1. Thus, values of a distribution function above 1 indicate more interevent times than predicted by the Poisson process, and conversely. Bottom panel is similar except that the synthetic seismicity includes only those events separated by a distance greater than the direct influence of the first event upon the hypocenter of the second. This plot indicates that events that do not border each other are nonetheless correlated in space and time; the correlation length of events is therefore larger than the average event size.

The exponential distribution (Poisson point process) has the unique property that the length of time since the last event provides no information about the probability of the next event. Distributions with more probability density at intermediate interevent times than the exponential and less probability at short and long interevent times occur in situations for which increasing interevent time increases the probability of an event. This situation would correspond to the "seismic gap" forecast model.[39] Physically, this can occur if all the earthquakes in a series rupture the same fault, and there is little stress transfer. If each earthquake relieves substantial stress on the fault, the next event will only occur after tectonic loading has a chance to bring the fault back up to failure stress levels. This is exactly the case, for example, for a single slider-block in which only events regularly spaced in time occur.

The slider-block data in Figure 19 (top) exhibit the opposite kind of behavior, similar to the natural seismicity. This is the clustering, or "anti-gap" forecast model,[23] in which an event is more likely to happen soon after a previous event, rather than later. Small values of interevent time ($< .25$) are more probable, and larger values ($.5$ to ~ 3.5) are less probable than the Poisson process would predict, implying the existence of clustering. Longer interevent times are also more probable than the Poisson point process, due to the fact that most of the activity is temporally concentrated within the clusters.

There are two major causes of clustering in seismicity, geometry of faults and dynamical interaction between faults. Fault geometries have been shown to be fractal (e.g., Brown and Scholz, Aviles et al.,[4] Turcotte,[71] and Sahimi et al.[58]) so the network of structures upon which earthquakes occur occupy space in an irregular and spatially clustered fashion. This mechanism for earthquake clustering is not present in the simpler slider-block models, because they represent a single planer fault. The second primary cause of clustering is stress transfer, in which a fault alters the stress state near it, and brings neighboring faults closer to failure. Stress transfer between faults on a large scale is an active area of research (e.g., Stein et al.,[70] Jaume and Sykes,[22] and Harris and Simpson[17]).

Increasingly convincing evidence has been compiled showing that changes in the static stress field are an important triggering mechanism beyond the usual aftershock zone, implying the existence of long-range correlations. A possible example of the existence of these long-range correlations has been given by Hill et al.,[20] which documents observations of the Landers earthquake triggering seismicity at very great distances, a conclusion that remains controversial. It has been argued that these long-range interactions must be due to the transient stresses that result from the passage of seismic waves, because static stress changes at great distances are so small. The nearest-neighbor slider-block model provides a convenient arena in which to test for correlations beyond the range of direct elastic effects. We now eliminate all pairs of events that border each other (nearest-neighbor events) and plot the results in Figure 19 (bottom). This leaves a distribution of interevent times for events which can have no direct interactions, and yet, the clustering remains. The degree of clustering is very similar to the Landers aftershock set shown on both plots, and cannot be attributed to dynamic effects either, because these are absent from the slider-block model. Clustering arises because the stress field has spatial and temporal correlations, extending beyond the range of direct stress transfer. The pattern of stress transfer has been reduced to an essential minimum, nearest-neighbor interaction, yet these models still show temporal clustering.

As summarized earlier (e.g., Rundle et al.[54]), slip evolution in the slider-block models induce a self-organization, which is in turn associated with a correlation length that may be much larger than the short-range, nearest-neighbor interaction distance. Effects like these are a very common feature of nonlinear, interacting systems near a critical point (in particular, see the discussion on pages 5–10 of Ma[30]; also the discussion on pages 75–78 and 164–179 of Nicolis and Prigogine[38]). It should be noted that these long-range correlations are always associated with the existence of a scaling, or power law relation such as the Gutenberg-Richter magnitude frequency relation. Failure of individual blocks, which represent fluctuations about the average steady state,[45] can thus become correlated over much longer ranges than the interaction distance.

Interactions between events thus overcome their own very short range nature to carry the correlation between events beyond the spatial limits of the direct interactions. Something similar could well be operating in the real earth, so that an event on one fault induces small amounts of slip (or strain) on its neighboring faults, which in turn induce yet smaller slip (or strain) on their many neighboring faults, etc. If the number of successive neighboring faults grows rapidly enough (power law), the result is that the occurrence of events on faults at large distances apart may become correlated in time. The same mechanism allows events close together in space to become correlated over long time intervals.

5. SUMMARY AND IMPLICATIONS FOR PREDICTABILITY

A variety of simple techniques have been proposed for predicting earthquakes (e.g., Scholz,[63] and Bakun and McEvilly[6]) including slip-predictable, time-predictable, and simple event-recurrence ("Parkfield") plots. Essentially, all of these methods involve fitting the recurrence pattern of events with straight lines to predict the next event in the sequence. These methods implicitly assume that stress accumulates on a fault at a well-determined, uniform rate; that the static stress drop is nearly uniform over the fault during an earthquake; that the effects of stress redistribution to other faults, or to the various parts of the same fault plane during the earthquake are negligible compared to the stress that is lost; that the rupture patterns of earthquakes are dominated by the spatial variations of strength along the faults, producing "characteristic" earthquakes; and that the rupture mechanism (hypocenter and directivity) in a temporal sequence of "characteristic" earthquakes repeats nearly exactly. To examine the validity of these prediction approaches and implicit assumptions, we have examined a simple slider-block model that has been shown to reproduce many of the observed features of seismicity on real faults (e.g., Rundle and Klein[52,53]). This model uses simple nearest-neighbor springs with spring constant K_C, loading springs with spring constant K_L, a spatially uniform failure threshold σ^F, and a simple block jump rule at the time of failure.

Using this simple model, we have found, for example, that magnitude-frequency distributions are produced that are representative of magnitude-frequency distributions for natural faults; that a pattern of "characteristic" earthquakes appears in the simulations, the pattern persisting for a period of time only to be altered eventually into a new pattern of events; and that space-time clustering is an important property of the simulations. Details of these results depend significantly on the parameters used in the simulations. It is known, for example, that requiring a fixed stress drop $\Delta\sigma_i = \sigma^F - \sigma^R$ during block slip leads to strictly periodic events.[14] Therefore, models with variable stress drop $\Delta\sigma_i = \sigma_i - \sigma^R$ and weak coupling (small R^2) can be expected to display approximate periodicity over limited time periods.

Our major result is that large events can be "predicted" only over limited time periods, the length of which depends on details of model parameters in the simulations. At the end of these limited periods, a dynamical reorganization occurs which leads to the appearance of a new but still temporary pattern of events. The nature of the processes responsible for this temporal reorganization is not as yet understood. The lack of strict periodicity is reflected both by changes in the temporal sequence, as well as ongoing changes in details of the temporal sequence of failing blocks during the synthetic earthquake. To the extent that the slider-block simulations capture important aspects of natural seismicity, it seems likely that natural earthquakes will be characterized by similar limits on predictability. In turn, this suggests that a deeper understanding of the physics of earthquake simulations will lead to fresh approaches for understanding natural events.

ACKNOWLEDGEMENTS

JBR would like to acknowledge support from NASA contract NAG5-2353 to the Cooperative Institute for Research in Environmental Sciences at the University of Colorado in Boulder, Colorado. The authors would also like to acknowledge a valuable critical review by J. R. Rice on an earlier version of the manuscript.

REFERENCES

1. Allen, M. P., and D. J. Tildsley. *Computer Simulations of Liquids.* Oxford: Clarendon Press, 1987.
2. Amit, D. J. *Field Theory, the Renormalization Group, and Critical Phenomena.* Philadelphia: World Scientific, 1984.
3. Arrowsmith and Place. *An Introduction to Dynamical Systems.* Cambridge, MA: Cambridge University Press, 1991.
4. Aviles, C. A., C. H. Scholz, and J. Boatwright. "Fractal Analysis Applied to Characteristic Segments of the San Andreas Fault." *J. Geophys. Res.* **92** (1987): 331.
5. Bak, P., and C. Tang. "Earthquakes as a Self-Organized Critical Phenomenon." *J. Geophys. Res.* **94** (1989): 15635–15638.
6. Bakun, W., and T. V. McEvilly. "Recurrence Models and Parkfield, California, Earthquakes." *J. Geophys. Res.* **89** (1984): 3051–3058.
7. Ben-Zion, Y., and J. R. Rice. "Earthquake Failure Sequences Along a Cellular Fault Zone in a 3D Elastic Solid Containing Asperity and Non-asperity Regions." *J. Geophys. Res.* **98** (1993): 14109–14131.
8. Binder, K. *Monte Carlo Methods in Statistical Physics.* Topics in Current Physics, vol. 7. Berlin: Springer-Verlag, 1986.
9. Binder, K. *Applications of the Monte Carlo Method in Statistical Mechanics.* Topics in Current Physics, vol. 36. Berlin: Springer-Verlag, 1987.
10. Brown, S. R., C. H. Scholz, and J. B. Rundle. "A Simplified Spring-Block Model of Earthquakes." *Geophys. Res. Lett.* **18** (1991): 215.
11. Burridge and Knopoff. "Model and Theoretical Seismicity." *Bull. Seism. Soc. Am.* **57** (1967): 341–371.
12. Carlson, J. M., and J. S. Langer. "Properties of Earthquakes Generated by Fault Dynamics." *Phys. Rev. Lett.* **62** (1989): 2632–2635.
13. Feynman, R. P., and A. R. Hibbs. *Quantum Mechanics and Path Integrals.* New York: McGraw-Hill, 1965.
14. Gabrielov, A., W. Newman, and L. Knopoff. "Lattice Models of Failure: Sensitivity to the Local Dynamics." *Phys. Rev. E* **50** (1994): 188–197.
15. Gross, S. J., and C. Kisslinger. "Tests of Models of Aftershock Rate Decay." *Bull. Seism. Soc. Am.* **84** (1994): 1571–1579.
16. Haberman, R. E. "Consistency of Teleseismic Reporting Since 1963." *Bull. Seism. Soc. Am.* **72** (1982): 93–111.
17. Harris, R. A., and R. W. Simpson. "Changes in Static Stress on Southern California Faults After the 1992 Landers Earthquake." *Nature* **360** (1992): 251.
18. Hauksson, E., L. M. Jones, K. Hutton, and D. Eberhart-Phillips. "The 1992 Landers Earthquake Sequence: Seismological Observations." *J. Geophys. Res.* **98** (1993): 19835–19858.

19. Heaton, T. "Evidence for and Implications of Self-Healing Pulses of Slip in Earthquake Rupture." *Phys. Earth & Planet. Int.* **64** (1990): 1–20.
20. Hill, D. P., P. A. Reasenberg, and A. Michael. "Seismicity Remotely Triggered by the Magnitude 7.3 Landers, California, Earthquake." *Science* **260** (1993): 1617.
21. Hopfield, J. J. "Neurons, Dynamics and Computation." *Physics Today* **47** (1994): 40–46.
22. Jaume, S. C., and L. R. Sykes. "Changes in State of Stress on the Southern San Andreas Fault Resulting from the California Earthquake Sequence of April to June 1992." *Science* **258** (1992): 1325–1328.
23. Kagan, Y. Y., and D. D. Jackson. "Long-Term Earthquake Clustering." *Geophys. J. Int.* **104** (1991): 117–133.
24. Kanamori, H., and Anderson. "Theoretical Basis for Some Empirical Relations in Seismelogy." *Bull. Seism. Soc. Am.* **65** (1975): 1073–1095.
25. Kanamori, H. "Global Seismicity." In *Earthquakes: Observation, Theory and Interpretation*, Course LXXXV, 596–608. Proc. Intl. School of Physics "Enrico Fermi," the Italian Physical Society. Amsterdam: North Holland, 1983.
26. Kanamori, H. "Mechanics of Earthquakes." *Ann. Rev. Earth Planet. Sci.* (1993): in press.
27. King, G. C. P., and J. Nabelek. "Role of Fault Bends in the Initiation and Termination of Earthquake Rupture." *Science* **228** (1985): 984–987.
28. Kostrov, B. V. "Self-Similar Problems of Propagation of Shear Cracks." *J. Appl. Mech.* **28** (1964): 1077–1087.
29. Lay, T., H. Kanamori, and L. Ruff. "The Asperity Model and the Nature of Large Subduction Zone Earthquakes." *Earthquakes Pred. Res.* **1** (1982): 3–71.
30. Ma, S.-K. *Modern Theory of Critical Phenomena.* Reading, MA: Benjamin-Cummings, 1976.
31. Ma, S.-K. *Statistical Mechanics* Philadelphia, PA: World Scientific, 1985.
32. McNally, K. C. "Plate Subduction of Earthquakes Along the Middle America Trench." In *Earthquake Prediction, An International Review.* AGU monograph 4. Washington, DC: American Geophysical Union, 1981.
33. Monette, L., and W. Klein. "Spinodal Nucleation as a Coalescence Process." *Phys. Rev. Lett.* **68** (1992): 2336–2339.
34. Mouritsen, O. G. In *Computer Studies of Phase Transitions and Critical Phenomena.* Berlin: Springer-Verlag, 1984.
35. Nakanishi, H. "Cellular Automaton Model of Earthquakes with Deterministic Dynamics." *Phys. Rev. A* **43** (1990): 6613–6621.
36. Nakanishi, H. "Cellular Automaton Model of Earthquakes with Deterministic Dynamics." *Phys. Rev. A* **43** (1990): 6613–6621.
37. Narkounskaia, G., J. Huang, and D. L. Turcotte. "Chaotic and Self-Organized Critical Behavior of a Generalized Slider Block Model." *J. Stat. Phys.* **67** (1992): 1151–1183.
38. Nicolis, G., and I. Priogine. *Exploring Complexity.* New York: W. H. Freeman, 1989.

39. Nishenko, S. P., and R. Buland. "A Generic Recurrence Interval Distribution for Earthquake Forecasting." *Bull. Seism. Soc. Am.* **77** (1987): 1382–1399.

40. Olami, Z., H. J. S. Feder, and K. Christensen. "Self-Organized Criticality in a Continuous, Non-Consecutive Cellular Automaton Modeling Earthquakes." *Phys. Rev. Lett.* **68** (1992): 1244–1247. (Comment by W. Klein and J. Rundle, *Phys. Rev. Lett.* **71** (1993): 1288–1289.)

41. Pacheco, J. F., C. H. Sholz, and L. Sykes. "Changes in Frequency-Size Relationship from Small to Large Earthquakes." *Nature* **355** (1982): 71–73.

42. Press, W. H., B. P. Flannery, S. A. Teukolsky, and W. T. Vetterling. *Numerical Recipes.* Cambridge: Cambridge University Press, 1988.

43. Richter, C. F. *Elementary Seismology.* San Francisco: W. H. Freeman, 1958.

44. Rundle, J. B., and Jackson. "Numerical Simulation of Earthquake Sequences." *Bull. Seism. Soc. Am.* **67** (1977): 1363–1378.

45. Rundle, J. B. "A Physical Model for Earthquakes: 1. Fluctuations and Interactions." *J. Geophys. Res.* **93** (1988): 6237–6254.

46. Rundle, J. B., and W. Klein. "Nonclassical Nucleation and Growth of Cohesive Tensile Cracks." *Phys. Rev. Lett.* **63** (1989): 171–174.

47. Rundle, J. B. "A Physical Model for Earthquakes: 3. Thermodynamical Approach and Its Relation to Nonclassical Theories of Nucleation." *J. Geophys. Res.* **94** (1989): 2839–2855.

48. Rundle, J. B., and W. Klein. "Nonlinear Dynamical Models for Earthquakes and Frictional Sliding: An Overview." In *Proc. 33rd Symposium Rock Mech.* Rotterdam: A. A. Balkema, 1992.

49. Rundle, J. B. "Magnitude Frequency Relations for Earthquakes Using a Statistical Mechanical Approach." *J. Geophys. Res.* **98** (1993): 21943–21949.

50. Rundle, J. B., and W. Klein. "Scaling and Critical Phenomena in a Class of Slider Block Cellular Automaton Models for Earthquakes." *J. Stat. Phys.* **72** (1993): 405–412.

51. Rundle, J. B., and D. L. Turcotte. "New Directions in Theretical Studies of Tectonic Deformation: A Survey of Recent Progress." In *Contributions of Space Geodesy to Geodynamics: Crustal Dynamics*, edited by D. E. Smith and D. L. Turcotte. Geodynamics Series, Vol. 23. Washington, DC: AGU, 1993.

52. Rundle, J. B., and W. Klein. "Dynamical Segmentation and Rupture Patterns in a 'Toy' Slider Block Model for Earthquakes." *Nonlin. Proc. Geophys.* **2** (1995): 61–81.

53. Rundle, J. B., and W. Klein. "New Ideas About the Physics of Earthquakes." *Rev. Geophys. Sp. Phys. Suppl.* (1995): 283–286.

54. Rundle, J. B., J. Ross, G. Narkounskaia, and W. Klein. "Earthquakes, Self-Organization, and Critical Phenomena." *J. Stat. Phys.* (1994): submitted.

55. Rundle, J. B., A. V. M. Herz, and J. Hopfield. Manuscript in preparation, 1994.

56. Rundle, J. B., W. Klein, S. Gross, and D. L. Turcotte. "Observation of Boltzman Fluctuations in Stochastic, Massless Slider-Block Simulations." This volume.

57. Sahimi, M., M. C. Robertson, and C. G. Sammis. "Relation Between the Earthquake Statistics and Fault Patterns, and Fractals and Percolation." *Physica A* **191** (1992): 57.

58. Sahimi, M., M. C. Robertson, and C. G. Sammis. "Fractal Distribution of Earthquake Hypocenters and Its Relation to Fault Patterns and Percolation." *Phys. Rev. Lett.* **70** (1993): 2186–2189.

59. Schwartz and Coppersmith. "Fault Behavior and Characteristic Earthquakes: Examples from the Wasatch and San Andreas Faults." *J. Geophys. Res.* **89** (1984): 5681–5698.

60. Smale. "Differentiable Dynamical Systems." *Bull. AMS* **73** (1967): 747–817.

61. Savage, J. C. "Criticism of Some Earthquake Forecasts of the National Earthquake Prediction Evaluation Council." *Bull. Seism. Soc. Am.* **81** (1991): 862–881.

62. Savage, J. C. "Empirical Earthquake Probabilities from Observed Recurrence Intervals." *Bull. Seism. Soc. Am.* **84** (1994): 219–221.

63. Scholz, C. H. *The Mechanics of Earthquakes and Faulting.* Cambridge: Cambridge University Press, 1990.

64. Schulman, L. S. *Techniques and Applications of Path Integration.* New York: John Wiley, 1981.

65. Schwartz, D. P., and K. J. Coppersmith. "Fault Behavior and Characteristic Earthquakes: Examples from the Wasatch and San Andreas Faults." *J. Geophys. Res.* **89** (1984): 5681–5698.

66. Schwartz, S. Y., J. W. Dewey, and T. Lay. "Influence of Fault Plane Heterogeneity on the Seismic Behavior in the Southern Kurile Islands Arc." *J. Geophys. Res.* **94** (1989): 5637–5649.

67. Senatorski, P. "Fault Zone Dynamics Evolution Patterns." Ph.D. Dissertation, Geophysical Institute of the Polish Academy of Sciences (Polska Akademia Nauk, Instytut Geofizyki), Warsaw, Poland, 1993.

68. Sieh, K. E., M. Stuiver, and D. Brillinger. "A More Precise Chronology of Earthquakes Produced by the San Andreas Fault in Southern California." *J. Geophys. Res.* **94** (1989): 603–623.

69. Stauffer, D., and A. Aharony. *Introduction to Percolation Theory.* London: Taylor and Francis, 1992.

70. Stein, R. S., C. P. King, and J. Lin. "Change in Failure Stress on the Southern San Andreas Fault System Caused by the 1992 Magnitude = 7.4 Landers Earthquake." *Science* **258** (1992): 1328–1332.

71. Turcotte, D. L. "A Fractal Approach to Probabilistic Seismic Hazard Assessment." *Tectonophysics* **167** (1992): 171.

72. Yeomans, J. M. *Statistical Mechanics of Phase Transitions.* Oxford: Clarendon Press, 1992.

Andreas V. M. Herz
Beckman Institute, University of Illinois at Urbana-Champaign, Urbana, IL 61801

A Comparison of Simple Earthquake Models: Self-Organized Criticality Versus Intermittent Phase Locking

Arrays of interconnected blocks with stick-slip friction play an important role as simple models of earthquake faults. This article focuses on some models with closely related microscopic dynamics but highly distinct collective phenomena. It is shown that one class of models exhibits rapid (intermittent) phase locking. This behavior is compared with numerical results from other models that indicate self-organized criticality. Implications for earthquake modeling and earthquake prediction are discussed.

INTRODUCTION

Earthquakes occur predominantly on fault systems located at the boundaries between tectonic plates. Relative movement of the plates generates stress that is slowly accumulated before it is quickly released during earthquakes. These earthquakes exhibit two most astonishing regularities. The first one concerns the frequency of earthquakes as a function of their size. Averaged over an entire fault system such as the San Andreas fault and its side branches, earthquakes occur at a rate that depends in a power-law fashion on the released energy E_0.[20] This relation is usually formulated in terms of accumulated energies, $p(E_0 > E) \propto E^B$. The "Gutenberg-Richter exponent" B depends on the fault system under consideration and varies between 0.3–0.9. The scaling relation holds over many orders of magnitude and suggests that fault systems exhibit critical behavior.

The second main feature of earthquake activity is the existence of seismic cycles. On some faults, the largest earthquakes (the so-called "characteristic events") have repeated with remarkable similarity from one recurrence to another. In certain cases, for example the Parkfield section of the San Andreas fault, recorded characteristic earthquakes had the same epicenter, magnitude, rupture direction and even identical foreshocks.[3] On the other hand, periodic activity is never exact due to the interaction of neighboring faults, nor does it last for more than a few cycles; on the Parkfield section, characteristic events (magnitude 5.5–6) occured in 1857, 1881, 1901, 1922, 1934, and 1966; and the next large earthquake, predicted for 1988 ± 7 has not happened yet.

Modeling the microscopic processes that eventually generate earthquakes is a formidable task. Earthquake faults are spatially extended systems, rock fracture involves complex dynamics with nonlinear frictional forces, and both the time course of "healing" processes in previously active regions and inhomogeneities in the rock material influence the system behavior. Thus analytically tractable models cannot be more than caricatures of the real phenomena, and even elaborate numerical studies of fairly detailed models remain highly controversial.[41]

Instead of attacking the original problem, one may thus opt for studying simple physical systems that share important details with the microscopic physics of earthquake processes. Promising candidates for such analogous model systems have been proposed by Burridge and Knopoff[8] who considered "slider-block" models, coupled systems of blocks and springs that are slowly pulled accross a rough surface. Numerous studies have revealed that for a variety of model parameters, systems of this type approach a critical state with statistical properties that are reminiscent of the Gutenberg-Richter power law.[8,10,11,42] However, the complexity of the dynamical equations has prevented a detailed analytical treatment or large-scale numerical simulations.

Further simplifications of the models are based on a separation of time scales. Earthquakes are rapid events that typically last for less than a minute. The stress-loading process, on the other hand, defines a time scale that is more than six orders of magnitude longer—the shortest recurrence times between characteristic events are in the ten-year range. Neglecting aftershocks, earthquakes may thus be approximated as *instantaneous* events, separated by silent episodes of gradual stress increase. The absence of a short time scale implies that such models cannot capture velocity-dependent friction. Stick-slip friction has to be incorporated by a static fracture criterion and one obtains a description in terms of massless pulse-coupled threshold elements.

Self-organized criticality has been found in some of these systems, both in discrete stochastic (sandpile-like) scenarios[1,2,14,25,52] and continuous deterministic models.[13,24,37,39] Although the paradigm of self-organized criticality may thus provide a heuristic explanation of the Gutenberg-Richter law in terms of simple microscopic systems, these models fail to exhibit characteristic earthquakes[24]: "In particular the existence of limit cycles with the appearance of earthquakes of a characteristic size for a given fault system...indicates that self-organized criticality is at best a partial description of the earthquake phenomena."[44]

In addition, there is considerable evidence that the Gutenberg-Richter law does *not* apply to single faults, only to entire fault systems.[30] "Many seismologists believe that each fault is only capable of generating earthquakes within a narrow magnitude range, related in a simple way to the linear dimensions of the fault (i.e., characteristic events). A variant of this opinion is one which contends that any given fault can generate either very small events or characteristic events."[32] An explanation of the observed scaling laws in terms of single-fault models may thus be a correct (and fascinating!) answer to a wrong question: "Although the power-law distribution for the rate of energy released describes seismicity on a broad two-dimensional network of faults, it is rare, if ever, that the power-law energy-rate distribution has been observed for an individual fault segment."[26]

If one follows the above assumption that the size of characteristic events scales with the fault size, then the observed power-law behavior is just a manifestation of the fractal geometry of fault systems.[46,47] This view does not provide any dynamical explanation of the Gutenberg-Richter law, but it shifts the attention from the time evolution of single faults to the generation and dynamics of entire fault systems.

The periodic recurrence of earthquakes has been modeled by using single driven threshold elements representing the mean stress level of an earthquake fault.[3,6,49] Since a single discharge corresponds to a system-wide earthquake, such scalar models cannot reproduce the spatial complexity of earthquakes. Spatially extended systems with site-dependent thresholds,[5,26,28,34,43,51] on the other hand, have been shown to exhibit seismic cycles with characteristic earthquakes and smaller events. The results are in agreement with the idea that periodic earthquake activity is due to regions with increased rock strength, so-called asperities, where the eventual slip zone of large earthquakes remains locked during the stress-accumulation phase of the seismic cycle.

The present article proves that seismic cycles can emerge as *collective* phenomena in *homogeneous* systems.[21] This shows that contrary to the general belief, inhomogeneities such as asperities are *not* required to generate long-term spatial localization of earthquakes. The systems used to advocate this purely dynamical explanation of periodic earthquake activity are strikingly similar to the model of Olami, Feder, and Christensen that exhibits self-organized criticality.[39] The similarity between both approaches demonstrates that even closely related microscopic descriptions of geophysical processes may lead to highly distinct emergent properties. Implications of this observation are discussed towards the end of the chapter.

THE MODEL SYSTEM

The model focuses on the dynamics of a single fault, represented at first by a rectangular two-dimensional lattice. Extensions are treated in a later section. The accumulated stress at site i is modeled by a scalar variable F_i which increases with unit velocity between "earthquakes." Assuming a separation of time scales as discussed in the Introduction, stick-slip friction is incorporated by a threshold process. When one of the F_i reaches the threshold F_{th}, it is immediately reset to zero. At the same time, the stresses F_j of i's four nearest neighbors (nn) are increased by $\alpha < F_{th}/4$. The rules provide an exact description of single slips in slider-block systems[8] as has been noted by Olami et al.[39]

Events with multiple slips occur if due to an elementary relaxation, a second site j becomes unstable, $F_j \geq F_{th}$, due to and "during" the slip of block i. In the present prescription it is reset to $\gamma(F_j - F_{th})$, $0 \leq \gamma \leq 1$, and again, the quantity α is transmitted to all neighboring sites.[21] For $\gamma = 1$, the stress drop equals the physical limit F_{th}; for $\gamma = 0$, slipping blocks are relaxed to the stress-free position $F_j = 0$. Both scenarios approximate the true relaxation process which cannot be incorporated due to the lack of a short time scale and inertial terms. If more than one site becomes unstable, either a fixed update order is chosen or the site with the largest stress is relaxed first.

Without loss of generality, F_{th} is set to unity. The system is initialized with random values for the F_i, independently drawn from a uniform probability distribution with width w such that $F_i \in [1 - w, 1]$. In a general context, one may consider systems with relaxations to a positive final stress value. Such systems can be mapped onto the present models if one allows $w > 1$. Thus this case will be discussed as well.

The time evolution can be summarized by the following rules:

i.   Initialize the F_i randomly in $[1 - w, 1]$.

ii. If $F_i \geq 1$ and if i is next in the update scheme, then

$$F_i \to F_i' = \gamma(F_i - 1) \tag{1}$$

and

$$F_{nn} \to F_{nn}' = F_{nn} + \alpha . \tag{2}$$

iii. Repeat step (ii) until $F_i < 1$ for all i.

iv. If the condition of step (ii) does not apply, then

$$dF_i/dt = 1 \quad \text{for all } i . \tag{3}$$

Except from the initial conditions, the rules are deterministic and thus differ from approaches motivated by sandpile models. The two limiting cases, $\gamma = 0$ and

$\gamma = 1$, are of particular interest. For $\gamma = 1$, the specific update order chosen in (iii) does not influence the evolution of the system in terms of its stable configurations (all $F_i < 1$). In addition, instantaneous updates (1) and (2) and slow dynamics (3) commute. The system is Abelian and equivalent to the avalanche model proposed by Gabrielov.[17]

For $\gamma = 0$, the earthquake model of Feder and Feder[16] is recovered in the limit $\alpha = 1/4$ and open boundary conditions. For general α, the present systems are almost identical with the model of Olami et al.[39] The latter is obtained if Eq. (2) is replaced by

$$F_{nn} \rightarrow F'_{nn} = F_{nn} + \alpha F_i .\tag{4}$$

A comparison with Eq. (2) shows that both models generate equal dynamics if only one site at a time is relaxed. Events with multiple slips, however, differ and give rise to distinct collective properties.

In the case of the model proposed by Olami et al., a *fixed fraction* of the released stress is transmitted to neighboring sites. In the present description, a *fixed amount* of stress is transmitted, independent of the initial position of the failed block. The physical motivation for this rule is that in a real system, blocks will always start their relaxation at a fixed failure threshold. If healing of a brokenmodels, slider-block bond occurs quickly, a failed block may accumulate the remaining stress from the slip-inducing process ($\gamma = 1$); otherwise the block remains at the stress-free position ($\gamma = 0$). Notice, in this context, that both the present model and that of Olami et al. implicitly assume that broken bonds heal before neighboring sites are updated. This assumption may be debatable from a geophysical point of view. Alternative approaches have been considered by various authors.[30,34,51]

THEORETICAL ANALYSIS

For periodic boundary conditions, the Abelian model ($\gamma = 1$) approaches limit cycles with period $P = 1 - 4\alpha$.[17,23] In what follows, it is shown that the attractors are reached as quickly as possible—as soon as every site has toppled once—and that this result holds for all $0 \leq \gamma \leq 1$.[21]

THEOREM 1. For periodic boundary conditions, and parameter values $\alpha < 1/4$ and $0 \leq \gamma \leq 1$, the dynamics generated by Eqs. (1)–(3) converge in finite time to a cyclic attractor with period P. The attractor is reached as soon as every block has slipped once. On the attractor, every site fails exactly once in a period.

Remark. Depending on the initial condition, the limit cycle can be very complex, and contain events in which one or a few blocks fail, and others in which many fail in synchrony, with the largest of these events corresponding to characteristic events (Figure 1).

PROOF. Any site i topples at most once during an earthquake. To cause a second failure, at least one neighbor would have to discharge twice before the ith element does so. By induction, this is impossible because $\alpha < 1/4$.

The stress increase due to toppling neighbors does not depend on whether those cells relax in a single or several distinct events. This implies that during any time interval of length P, an arbitrary site i relaxes no more than once: F_i increases by at most $1 - 4\alpha$ due to Eq. (3) and by up to 4α if all neighbors topple. At least the same total amount is lost in a single relaxation Eq. (1).

It is next shown that all transients have a finite duration. Let $t_{\max}$ denote the first time where all sites have failed at least once, t_i the last instance where site i topples before $t_{\max}$, $t_{\min}$ the minimum of all t_i, and j a site that failed at $t_{\min}$ without being triggered by other cells at that instant. By definition, all sites discharge at least once in $[t_{\min}, t_{\max}]$. This means in particular that every neighbor of j slips at least once between $t_{\min}$ and $t_{\max}$. Each event adds α to F_j. By assumption, site j fails at most once in $[t_{\min}, t_{\max}]$ so that $t_{\max} - t_{\min} < P$. The result implies that every site fails exactly once in $[t_{\min}, t_{\max}]$ and no site fails in $(t_{\max}, t_{\min} + P)$. Since $t_{\max} \leq w$, this proves that in finite time $t_{\min}$, all limit cycles are approached in the sense that $F_i(t) = F_i(t + P)$ for all $t_{\min}$. The argument also shows that the attractors are reached as soon as every element has slipped once. This finishes the proof. ∎

Notice that the update order in (iii) does not enter the proof and that one could also replace Eq. (1) by *any* rule for which $F_i - 1 \geq F_i' \geq 0$. Perhaps surprisingly, this includes stochastic update rules. For $\gamma < 1$, limit cycles with period P and one slip per cycle (and site) cannot occur if a site is driven above threshold because a single stress drop would be larger than unity, the total stress increase over one cycle. This means that the system has to relax to fine-tuned states where if site i is triggered by n of its neighbors at time t, $F_i(t^-) = 1 - n\alpha$. (P-periodic solutions of the model of Olami et al. have to satisfy the same condition.) It follows that although every toppling sequence of the Abelian model can be realized for arbitrary $\gamma < 1$, the volume of all attractors is greatly reduced when measured in the space of the dynamical variables F_i.

In the limit $\alpha \to 0$, rules (2) and (4) become identical. This may explain why the model of Olami et al.[39] quickly freezes in a periodic state for small to intermediate α.[19]

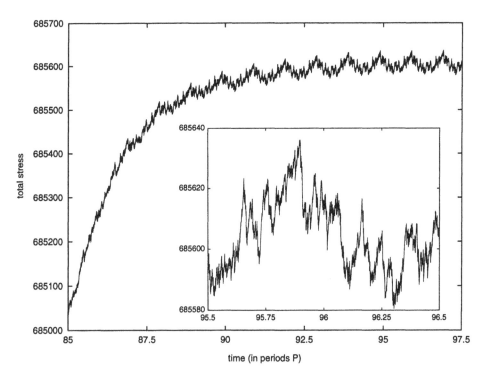

FIGURE 1   Transient and periodic limit cycle. Time evolution of the total stress $F_{\text{total}} = \sum_i F_i$ in a system with 1024^2 sites and periodic boundary conditions; $\gamma = 1$, $w = 1$, and $\alpha = 0.2495$. Vertical drops of size Δ correspond to events with $P^{-1}\Delta = 500\Delta$ slipped sites.

The limiting case $\gamma = 0$ allows a second, unrelated convergence proof.[23] A similar proof has independently been given by Gabrielov et al.[18] Consider the total accumulated stress,

$$F_{\text{total}} = \sum_i F_i \ . \tag{5}$$

THEOREM 2. Assume that the conditions of Theorem 1 hold and that $\gamma = 0$. Then

$$F_{\text{total}}(t + P) \geq F_{\text{total}}(t) \ . \tag{6}$$

PROOF. The change of F_{total} in the interval P is

$$F_{\text{total}}(t+P) - F_{\text{total}}(t) = (1 - 4\alpha)[N - \sum_i n_i(t, t + P)] \tag{7}$$

where $n_i(a, b)$ denotes the number of slips of site i in the time interval $[a, b)$. Since $n_i(t, t+P) < 1$ as shown before, the change of F_{total} in each time interval P is nonnegative. Because F_{total} is also bounded, it is a Lyapunov function: the system performs an "uphill march" in the landscape generated by F_{total}— if time is measured in steps of fixed length P. For an illustration, see Figure 1. Independent from Theorem 1, the result implies that after a finite time each $F_i(t)$ is periodic with period P and that every neuron fires once in any interval of length P. ∎

To avoid the somewhat unfamiliar evaluation of the Lyapunov function F_{total} at the discrete times $t + kP$, one may alternatively use the functional $L = \int_0^P F_{\text{total}}(s)ds$. Along solutions of Eqs. (1)–(3), L is differentiable with $dL(t)/dt = F_{\text{total}}(t + P) - F_{\text{total}}(t)$ for all $t \geq 0$, and the same conclusions are reached.

Properties of the limit cycles are governed by the two independent parameters α and w, characterizing the dependence upon the internal dynamics (α) and initial conditions (w). For small α and large enough w, events are expected to be localized with exponential size-frequency relation as observed in numerical experiments (Figure 2). If, on the other hand, $w \leq \alpha$, the very first rupture has to lead to a system-wide avalanche, followed by a globally synchronized oscillation. Global synchronization does also occur if local stress variations are smoothed out during very long transients. This phenomenon has been observed for $\gamma = 0$ and $w \gg 1$.

If $w \leq 2\alpha$, a toppled row triggers the entire adjacent row as soon as a single element of the second row fails. The example illustrates domino effects due to stress accumulation. Phenomena of this kind are only partially exhibited by dynamical descriptions based on Leath algorithms[27] and severely question percolation models of seismic activity.[29,30] Percolation theory may, however, be used to derive necessary avalanche conditions. For example, if $w \leq p_c^{-1}\alpha$, the very first slip has to result in a percolating set of ruptured sites with probability one. Here, $p_c \approx 0.5927$ denotes the percolation threshold for random site percolation.

If $w + 4\alpha < 1$, a slipped site is reset to a value that is smaller than the minimal stress of blocks that have not yet failed. Since the system is driven uniformly, toppled sites cannot catch up with the remaining elements. Thus every cell slips exactly once during the transient. This implies that the entire time evolution depends only on the ratio $r = w/\alpha$ and that the toppling sequences do *not* depend on γ.

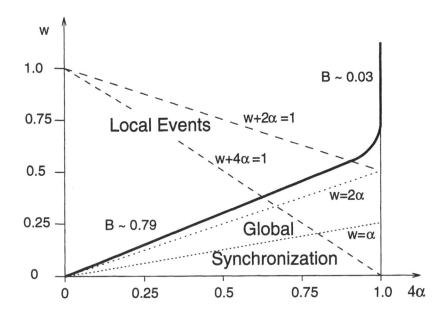

FIGURE 2 Dependence upon internal dynamics (α) and initial conditions (w) for periodic boundary conditions. Local events and global synchronization are separated by a line where the size-frequency relation obeys a power law. The solid line represents a numerical estimation for $\gamma = 1$. The line is valid for arbitrary $0 \leq \gamma \leq 1$ if $w + 2\alpha < 1$. The other, auxiliary lines refer to theoretical bounds discussed in the text.

Slips of yet unruptured areas have to occur if their corners reach threshold. At time t, the stress of a corner site is at least $1 - w + 2\alpha + t$, the stress of the block that failed first is at most $4\alpha + t$. This implies that for $w + 2\alpha < 1$, all sites slip before the first block slips twice. The previous bound $w + 4\alpha < 1$ can thus be improved to $w + 2\alpha < 1$.

To describe more general interactions, Eq. (2) is replaced by

$$F_j \to F_j' = F_j + T_{ji} \tag{8}$$

where T_{ji} denotes the amount of stress transmitted from i to j. Possible extensions include systems of arbitrary dimension and aspect ratio, and inhomogeneous, non-isotropic and long-range interactions to represent granularity or depth-dependent material properties. For nonnegative coupling strengths that satisfy the condition

$$\sum_i T_{ji} = A \leq 1 \quad \text{for all } j \tag{9}$$

the first convergence proof remains valid, with 4α replaced by A. (The same is true for the second proof if the couplings satisfy the additional constraint $\sum_j T_{ji} = A$.) Statistical properties, however, depend on the specific model assumptions.

To account for spatially varying loading I_i and nonseismic stress decrease due to slow creep, Eq. (3) could be replaced by a leaky-integrator equation

$$dF_i/dt = -F_i\tau^{-1} + I_i \ . \tag{10}$$

The equation points to a potentially rich connection between slider-block models and networks of integrate-and-fire neurons.[21,22,23,45]

NUMERICAL EXPERIMENTS

Numerical experiments were performed on square lattices with up to 1024^2 sites and nearest-neighbor interactions. The simulations were based on a variant of an algorithm proposed by Grassberger.[19] Data were taken from single limit cycles and multiple runs, and indicate that the size-frequency relation is self-averaging. Below the line $w + 2\alpha = 1$, and with γ set to unity the size-frequency relation follows a power law at $r = 2.44 \pm 0.02$. This value may coincide with $(1 - p_c)^{-1}$ and would then suggest that the collective properties of the present system are related to those of next-nearest-neighbor site percolation.

The energy released during an earthquake is given by the sum of over all stress changes. For $\gamma = 1$, the seismic moment is thus proportional to the number of slipped sites. For $\gamma \neq 1$, this is no longer true during the transient where a triggered block will release more energy than an initiator site. (This observation applies also to the model of Olami et al.)

Using the earthquake size as a measure for the released energy, a Gutenberg-Richter exponent $B = 0.79 \pm 0.05$ was obtained. Determined from sample averages of the first event after reaching the limit cycles only, it is $B_{\text{first}} = 0.05 \pm 0.05$. The difference is due to nonuniform stress accumulation during the lattice filling process and is also reflected in temporal inhomogeneities seen on the limit cycles. For $w \gg 1$ and $\alpha \to 1/4$, the transient times diverge when measured in terms of P, and $B = B_{\text{first}} = 0.03 \pm 0.03$.

The results reported so far have been obtained with periodic boundary conditions. For open boundary conditions, sites located at the edges (and corners) of the lattice receive less pull because they have fewer neighbors than sites in the bulk. Thus they cannot sustain the maximum failure rate P^{-1} and generate dynamical defects that propagate into the system and prevent complete phase locking. For $\gamma = 1$, the sizes of synchronized clusters gradually change due to the loss and

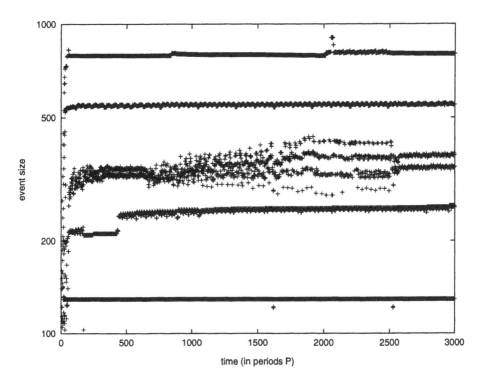

FIGURE 3 Intermittent behavior. Each cross marks the size ($\geq$ 10) and time of an event in a system with 50^2 sites and open boundary conditions; $\gamma = 0$, $w = 1$, and $\alpha = 0.2$. On a short time scale the system exhibits almost periodic behavior where events "repeat" after a time $T = (1 - 3\alpha) = 2P$.

new recruitment of phase-locked cells and one observes (quasi-) periodic behavior[17] with exponential size-frequency relations for $\alpha < 1/4$. For $\gamma = 0$, the fine-tuning described earlier leads to an intermittent behavior where clusters remain virtually unchanged over many cycles before they suddenly merge or break apart (Figure 3). The main slip frequency shifts from $(1 - 4\alpha)^{-1}$ to $(1 - 3\alpha)^{-1}$ as is demonstrated in Figure 4 and shows that the collective behavior of the entire system is strongly influenced by its boundary, a situation characteristic for extended driven systems.

Long transients and intermittency complicate quantitative numerical investigations. Elaborate simulations reveal, however, that for small γ, sample averages exhibit self-organized criticality. (Whether this is also true on the level of individual samples remains an interesting open question. It will be very hard to find an answer using numerical techniques because the durations of episodes with almost periodic behavior quickly grow with increasing system size.) The exponents

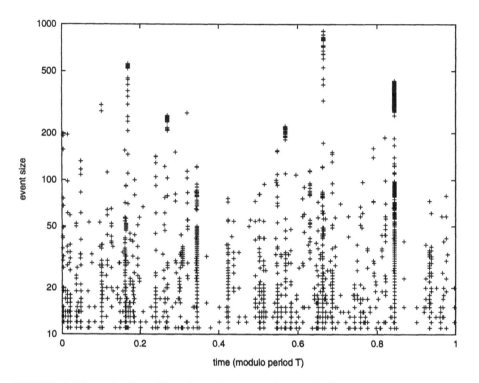

FIGURE 4   Phase locking. Plotted are the same data as in Figure 3, but time is now shown modulo T. Notice the extreme synchrony of events even during coalescence and disintegration of clusters.

depend on the level of stress conservation which is reminescent of the situation in the model of Olami et al. For $\gamma = 0$, it varies between $B = 0.40 \pm 0.05$ at $\alpha = 0.10$ and $B = 0.52 \pm 0.05$ (see also Feder and Feder[16]) for $\alpha \to 1/4$.

In the model of Olami et al., the time evolution of single systems is quite different (Figure 5). Although the dynamical rules are capable to generate large clusters of simultaneously slipping blocks[36,39,50]—a necessary requirement for power-law behavior—these clusters quickly disintegrate: since a *fixed fraction* of stress is transmitted, even small changes in the initial stress of a failed site may cause neighboring sites to topple out of synchrony and thus result in a break-up of the cluster. In systems where a *fixed amount* of stress is transmitted, this phenomenon cannot occur. Figure 6 shows the resulting differences in the size-frequency relations obtained from single realizations of the two model systems—self-organized

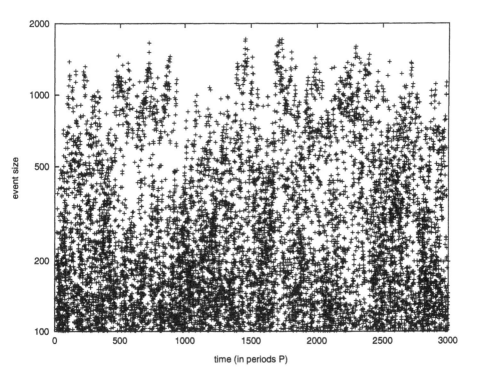

FIGURE 5   Time evolution in the model of Olami et al. Except from the changed redistribution rule, system parameters and dynamics are the same as in Figures 3 and 4.

criticality versus intermittent phase locking. The figure also demonstrates how the averaged size-freqency relation of the present model approaches power-law behavior with increasing number of samples.

DISCUSSION

The results presented in this article demonstrate that simple homogeneous earthquake models are capable to generate earthquake sequences with recurrent characteristic events. For periodic boundary conditions, all earthquakes repeat in a cyclic fashion once the attractor is reached. For the more realistic case of open boundary conditions, there are still remarkably long episodes with a nearly periodic recurrence

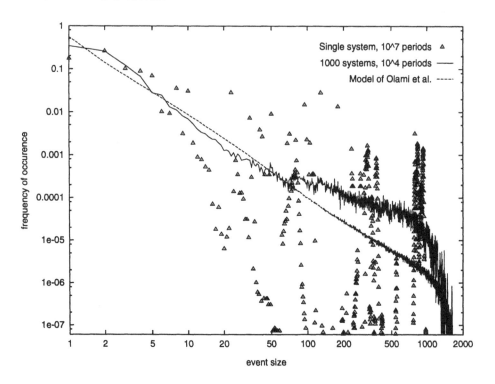

FIGURE 6 Size-frequency relations. Using the same parameters as in Figures 3–5, the frequency of occurence of an event is plotted as a function of the event size. Shown are results from a single simulation over 10^7 "periods" P (triangles), data from 1000 different runs, taken over 1000 periods after a transient of 5000 periods (solid line), and as a comparison, a single simulation of the model of Olami et al. over 10^5 periods (dashed line).

of characteristic events. This proves that unlike frequently stated,[26,38] neither inhomogeneities in the couplings between blocks nor complicated model dynamics are needed to assure long-time spatial localization of earthquakes. Surprisingly, this issue has not been discussed in the literature on Abelian models.[18]

The results are in accordance with the concern of some authors[5,34] that self-organized criticality should not be considered as *the* generic feature of simple fault models. In fact, this view may also apply to more realistic approaches to earthquake dynamics as exemplified by the following quote by Rice: "The numerical results show richly complex slip, with a spectrum of event sizes, when solved for a grid with oversized cells, that is, with cell size h that is too large to validly represent the underlying continuous system of equations. However, in every case for which it has been feasible to do the computations (moderately large L only), that spatiotemporally complex slip disappears in favor of simple limit cycles of periodically repeated large earthquakes with reduction of cell size h."[41]

The strong effects of boundary conditions on both the present model and that of Olami et al. show that earthquake models behave rather differently from systems known from solid-state physics such as the Ising model. This observation questions the extent to which methods developed for the latter class of systems can be applied to earthquake models.[44] Similarly, epidemic growth models or models based on percolation algorithms[29,31] fail to capture domino effects due to stress accumulation processes. Finally, it is obvious that a significant part of the analytical results obtained for conservative sandpile models cannot be applied to continuously driven, nonconservative systems. It thus seems that a completely new set of mathematical tools has to be developed to deal even with the most primitive earthquake models. Analytical studies such as those by Gabrielov,[17] Gabrielov et al.,[18] Middleton and Tang, and Socolar et al.[50] are important steps in that direction.

The comparison of different simple fault models shows that minute model details may have a significant influence on the collective system behavior—see also Gabrielov.[18] This situation is symptomatic for (over)-simplified models of complex systems. It raises doubts as to what role these models can play as quantitative descriptions of real earthquake processes. Statements such as "Models hold promise for predicting earthquakes" (Santa Fe Institute Bulletin[7]) appear premature in the light of these results. This is not to say that simple models may not become extremely important tools to develop and test earthquake-prediction algorithms under conditions that admit experimental falsification.[1,12,40,48]

The analytical and numerical results presented in this article confirm the observation that, indeed, "the statistical nature of the observed phenomena is strongly influenced by the loading, the form of the failure threshold and the mechanism for energy dissipation."[44] This applies in particular to the question of power-law behavior—"the presence or absence of a critical state depends delicately on the choice of the rules, and is very sensitive to small changes."[18] The results prove that simple unifying concepts—conceptually appealing as they may be—do not capture the richness observed both in nature and model systems.

ACKNOWLEDGMENTS

The results reported in this chapter were obtained in collaboration with John J. Hopfield. Our work was stimulated by a series of most helpful joint discussions with John Rundle. I would also like to thank the organizers of the "Workshop on Natural Hazards Reduction" for the invitation, the participants for interesting comments. I am grateful to Jean Carlson who contributed with valuable suggestions during two later visits to Santa Barbara, to Alan Middleton who made unpublished material available, and to Klaus Schulten who provided powerful computing facilities. This work has been supported by a Beckman Institute Fellowship and EC grant ERBCHBICT941486.

REFERENCES

1. Bak, P., and K. Chen. "Predicting Earthquakes." In *Nonlinear Structures in Physical Systems*, edited by L. Lam and H. C. Morris. New York: Springer-Verlag, 1989.
2. Bak, P., and C. Tang. *J. Geophys. Res.* **94** (1989): 15635.
3. Bakun, W. H., and A. G. Lindh. *Earth. Predict. Res.* **3** (1985): 285.
4. Barriere, B., and D. L. Turcotte. *Geophy. Res. Lett.* **18** (1991): 2011.
5. Brown, S. R., C. H. Scholz, and J. B. Rundle. *Geophy. Res. Lett.* **18** (1991): 215.
6. Bufe, C. G., P. W. Harsh, and R. O. Burford. *Geophys. Res. Lett.* **4** (1977): 91.
7. *The Bulletin of the Santa Fe Institute* **9** (1994): 17.
8. Burridge, R., and L. Knopoff. *Bull. Seism. Soc. Am.* **57** (1967): 341.
9. Carlson, J. M., and J. S. Langer. *Phys. Rev. Lett.* **62** (1989a): 2632.
10. Carlson, J. M., and J. S. Langer. *Phys. Rev. A* **40** (1989b): 6470.
11. Carlson, J. M., J. S. Langer, and B. E. Shaw. *Rev. Mod. Physics* **66** (1994): 657.
12. Christensen, K., and Z. Olami. *J. Geophys. Res.* **97** (1992): 8729.
13. Diaz-Guilera, A. *Phys. Rev. A* **45** (1992): 8551.
14. Ellsworth, W. L. *Nature* **363** (1993): 206.
15. Feder, H. J. S., and J. Feder, *Phys. Rev. Lett.* **66** (1991): 2669.
16. Gabrielov, A. *Physica A* **195** (1993): 253.
17. Gabrielov, A., W. I. Newman, and L. Knopoff. *Phys. Rev. E* **50** (1994): 188.
18. Grassberger, P. *Phys. Rev. E* **49** (1994): 2436.
19. Gutenberg, B., and C. F. Richter. *Ann. di Geofis.* **9** (1956): 1.
20. Herz, A. V. M., and J. J. Hopfield. To appear.
21. Hopfield, J.J. *Physics T oday* **47** (1994): 40.
22. Hopfield, J. J., and A. V. M. Herz. To appear.
23. Huang, J., G. Narkounskaia, and D. L. Turcotte. *Geophys. J. Int.* **111** (1992): 259.
24. Ito, K., and M. Matsuzaki. *J. Geophys. Res.* **95** (1990): 6853.
25. Knopoff, L., J. A. Landoni, and M. S. Abinante. *Phys. Rev. A* **46** (1992): 7445.
26. Leath, P. L. *Phys. Rev. B* **14** (1976): 5046.
27. Lomnitz-Adler, J. *Geophys. J. Roy. Astr. Soc.* **83** (1985): 435.
28. Lomnitz-Adler, J., and P. Lemus-Diaz. *Geophys. J. Int.* **99** (1989): 183.
29. Lomnitz-Adler, J. *Geophys. J. Int.* **108** (1992): 941.

30. Lomnitz-Adler, J., L. Knopoff, and G. Martinez-Mekler. *Phys. Rev. E* **45** (1992): 2211.
31. Lomnitz-Adler, J. *J. Geophys. Res.* **98** (1993): 17745.
32. Manna, S. S., L. B. Kiss, and J. Kertesz. *J. Stat. Phys.* **61** (1990): 923.
33. Matsuzaki, M., and H. Takayasu. *J. Geophys. Res.* **96** (1991): 19925.
34. Middleton, A. A., and C. Tang. To appear.
35. Nakanishi, H. *Phys. Rev. A* **41** (1990): 7086.
36. Newman, W. I., and L. Knopoff. *Geophys. Res. Lett.* **10** (1983): 305.
37. Olami, Z., H. J. S. Feder, and K. Christensen. *Phys. Rev. Lett.* **68** (1992): 1244.
38. Pepke, S. L., J. M. Carlson, and B. E. Shaw. *J. Geophys. Res.* **99** (1994): 6769.
39. Rice, J. R. *J. Geophys. Res.* **98** (1993): 9885.
40. Rundle, J. B., and D. D. Jackson. *Bull. Seism. Soc. Am.* **67** (1977): 1363.
41. Rundle, J. B., and S. R. Brown. *J. Stat. Phys.* **65** (1991): 403.
42. Rundle, J. B., and W. Klein. *J. Stat. Phys.* **72** (1993): 405.
43. Rundle, J. B., A. V. M. Herz, and J. J. Hopfield. To appear.
44. Sahimi, M., M. C. Robertson, and C. G. Sammis. *Physica A* **191** (1992): 57.
45. Sahimi, M., M. C. Robertson, and C. G. Sammis. *Phys. Rev. Lett.* **70** (1993): 2186.
46. Shaw, B. E., J. M. Carlson, and J. S. Langer. *J. Geophys. Res.* **97** (1992): 479.
47. Shimazaki, K., and T. Nakata. *Geophys. Res. Lett.* **7** (1980): 4.
48. Socolar, J. E. S., G. Grinstein, and C. Jayaprakash. *Phys. Rev. E* **47** (1993): 2366.
49. Takayasu, H., and M. Matsuzaki. *Phys. Lett. A* **131** (1988): 244
50. Zhang, Y. C. *Phys. Rev. Lett.* **63** (1989): 470.

W. Klein, * **C. Ferguson,** * **and John B. Rundle** **

*Department of Physics, and Center for Polymer Studies, Boston University, Boston MA 02215
**Department of Geological Sciences and, Cooperative Institute for Research in Environmental Sciences, University of Colorado, Boulder, CO 80309

Spinodals and Scaling in Slider-Block Models

Measurement of seismic energy released in earthquake events taken from faults worldwide has indicatated that there is scaling behavior (Gutenberg-Richter) associated with the earthquake process. In this chapter we will present evidence from simulations indicating that this scaling might be obtainable from properly chosen slider-block models and that it is associated in these models with a spinodal instability.

1. INTRODUCTION

Ever since Gutenberg and Richter[25] pointed out that the energy released in earthquake events scaled (i.e., the number of events with seismic moment M_o scaled as $M_o^{-2b/3}$ with $M_o \sim A^{3/2}$ where A is the area of the earthquake event and the exponent b is in the range $0.8 - 1.0$ for small events and $1.25 - 1.55$ for large events), geologists, geophysicists and physicists have been attempting to explain the origin of that scaling.

Reduction & Predictability of Natural Disasters, Eds. Rundle, Turcotte, & Klein,
SFI Studies in the Sciences of Complexity, Vol. XXV, Addison-Wesley, 1996

To any scientist familiar with advances in Statistical Mechanics over the last three decades, scaling plots are a strong indication of underlying long-range correlations and cooperative behavior, e.g., critical phenomena. If the Gutenberg-Richter scaling is indeed indicative of long-range correlations, this would provide a paradigm with which one could understand several aspects of earthquakes and might lead to some predictive capability.

However, earthquake phenomena are considerably more complicated than scaling plots alone indicate. In addition to scaling, there is space-time clustering of events and migration of activity along fault systems. Moreover, there are faults which appear to have events of roughly the same size that appear quasi-periodically in apparent contradiction to the idea that the scaling of events is the property of single faults. The latter phenomenon has led several researchers to propose that the scaling of earthquakes is a property only of the worldwide fault network.[25]

In order to study the statistical aspects of earthquake events,[2] Burridge and Knopoff approximately 30 years ago proposed a model that could be studied both theoretically and numerically and hopefully contained the essential physics of earthquake faults. The Burridge-Knopoff model consists of a network of blocks coupled by springs with force constants K_C sliding on a frictional surface. Each block is also attached to a loader plate via a spring with spring constant K_L. After the initial random placement of the blocks, the loader plate is moved with a velocity V, and the forces on the blocks computed. If the magnitude of the force vector ζ_i on block i is increased to the point where it exceeds a prescribed threshold value ζ_i^F, the block jumps or slides a distance U_i in the direction of the force, thereby reducing the force on that block to a residual value ζ^R. Each block that slides may induce failure of neighboring blocks by means of the coupling through the springs leading to clusters of failed blocks.

The dynamics of this model has been the subject of considerable interest, initially among seismologists[8,10,20,21] and more recently in the Condensed Matter Physics community.[1,3,4,6,23] For the most part the latter work has studied the behavior of massive blocks subject to a velocity-weakening friction force. Although cluster scaling appeared to be obtained from the study of the Burridge-Knopoff model, there has been no clear connection between the cluster scaling and any underlying critical phenomena. Attempts were made to describe earthquake events as fluctuations in a self-organized critical phenomena process,[1] but as we will show the situation is more complicated.

In order to address these problems we have studied a cellular automata version of the Burridge-Knopoff model introduced by Rundle and Brown.[21] The relative simplicity of the cellular automaton has allowed extensive simulations with wide variations of the parameters that specify the model. Our main result is that in the most realistic version of the model, the one that seems most likely to describe most earthquake faults, the data suggest that there is a spinodal or critical like point that underlies the scaling. In addition we can identify diverging time scales and fluctuations associated with the scaling. There is also some indication that quasi-periodic events can also occur and that the phenomena exhibited by the model is strongly dependent on the values assigned to the various defining parameters.

The remainder of this chapter is structured as follows. In Section 2 we describe the cellular automata model and discuss its justification. In Section 3 we describe the concept of the spinodal and how it appears in condensed matter physics. This description will concentrate on spinodals in Ising models. In Section 4 we describe our results; in Section 5 we discuss some physical reasons for our interpretation of the simulation data; and in Section 6 we summarize and state our conclusions.

2. CELLULAR AUTOMATON MODEL

We begin this section by defining the cellular automata version of the Burridge-Knopoff model. As with the Burridge-Knopoff model we consider blocks attached to each other via springs with spring constant K_C. The blocks are also attached via springs with spring constant K_L to a loader plate moving at a "velocity" V. All springs are to be considered as leaf springs. That is, they only exert a force in the direction of the plate motion or opposite to it. This restriction to a one-dimensional force is one of several simplifications of the Burridge-Knopoff model that facilitates simulation. After the positions of the blocks are set at random, the plate is moved a distance $V\Delta$ where Δ sets one time scale. The motion of the plate adds a spatially uniform increment of stress $VK_L\Delta$ to each block. After the plate is moved the blocks are checked to ascertain if their stress exceeds a prescribed failure threshold ζ_i. We have used a flat threshold ($\zeta_i^F = \zeta^F$) or a random threshold ($\zeta_i^F = \zeta^F + \epsilon$ where ϵ is a random number between two prescribed limits. If $\zeta_i < \zeta_i^F$, the block is moved a distance

$$J(\zeta_i) = \frac{\zeta_i - \zeta_i^R}{K} \tag{1}$$

where $K = K_L + qK_C$ and q is the number of neighboring blocks attached to the ith block by a spring and $\zeta_i^R < \zeta_i^F$ is the prescribed residual force.

The displacemet $U_i(t+1)$ at time $t+1$ is related to the displacement at time t by

$$U_i(t+1) = U_i(t) + J(\zeta_i)\Theta(\zeta_i - \zeta_i^F) \tag{2}$$

where $\Theta(\zeta_i - \zeta_i^F) = 0$ if $\zeta_i - \zeta^F < o$ and one otherwise. The stress on each block is

$$\zeta_i = \sum_j T_{ij}U_j(t) + Q_i(t) \tag{3}$$

where $T_{ij} = K_C$ for $i \neq j$ and $T_{ii} = -K$. The force from the loader plate

$$Q_i(t) = K_L V\Delta[\sum_n \Theta(n-t)] \tag{4}$$

The process of moving blocks if their stress exceeds ζ_i^F continues until all blocks have $\zeta_i < \zeta_i^F$. The plate is then moved a distance $V\Delta$ and the entire process repeats. Note that the loader plate applied stress $Q_i(t)$ is a function of V (Eq. (4)) so that small V results in a small stress increment due to a plate move.

Before presenting the results of our simulations, it will be useful to discuss two topics that are essential in interpreting our results. One, the nature of the spinodal, will be discussed in the next section. The other, the rationale behind the cellular automata model, we will discuss in the remainder of this section.

The cellular automata model is a version of the Burridge-Knopoff model. The Burridge-Knopoff model can be described by Newtons equations with a velocity weakened friction force. In the cellular automata we neglect the inertial term (i.e., mass times the acceleration) which is thought to be small compared to the friction term and the force from the springs. This is equivalent to the statement that any wave like motion in the slip will be strongly damped. This is discussed in detail in Rundle and Klein.[12]

The second important approximation in the cellular automata model is that the velocity-weakened friction force is replaced by a failure threshold so that the block slips only if the stress on the block exceeds the threshold. Moreover, the block slips an amount given by the jump rule of Eq. (1). This approximation follows from the fact that we neglect the inertial term and is also discussed in Rundle and Klein.[12]

The final aspect of the cellular automata model we will discuss is that in real faults the interactions between rock segments is taken from linear elasticity theory.[22] The stress Green's function $T(\vec{x} - \vec{x}')$ is then taken to be that of an elastic medium, i.e.,

$$T(\vec{x} - \vec{x}') \sim \frac{1}{|\vec{x} - \vec{x}'|^3} . \tag{5}$$

In the cellular automata model the spring constants are related to the moments of the stress Greens function. In particular[22]

$$K_L = -\int T(\vec{r})d\vec{r}, \tag{6}$$

and

$$qK_C = \frac{1}{2}\int T(\vec{r})|\vec{r}|^2 d\vec{r}, \tag{7}$$

where K_C is the spring constant for a single spring.

We have defined the second moment in this way to make it easier to describe the simulations. We will come back to this point in Section 4. For the moment we will simply note that the form of the stress Greens function in Eq. (5) implies that for faults, which are assumed to be $d = 2$, $q \to \infty$. This in turn implies that earthquake faults are described by mean-field theories. This is the point that will be taken up in the next section.

3. SPINODALS AND MEAN-FIELD THEORIES

The van der Waals theory of fluids and the Curie-Weiss theory of magnetism generally depict the metastable phase as an analytic continuation of the stable phase. For example in the van der Waals equation, sketched schematically in Figure 1, the stable phase is the solid line which includes the Maxwell construction, and the metastable phase (dashed line) is the analytic continuation of the stable phase. The maximum and minimum of the curve (marked with B and C) are the spinodals. They represent the limit of the metastable phase and are critical points. Obviously the isothermal compressibility κ_T defined as $\kappa_T = \frac{1}{V}[\frac{\partial P(\bar{V},T)}{\partial V}]^{-1}$ diverges at the spinodals as does the correlation length.[19] Here $\bar{V}$ is the volume.

In magnetic systems such as Ising models a similar structure exists. In Figure 2 we plot the magnetization as a function of the magnetic field. At the points marked by B and C the isothermal susceptibility $\chi_T = \frac{\partial m(T,h)}{\partial h}$ diverges, and the spinodals mark the limit of the metastable phase (see Figure 2).

Although the van der Waals and Curie-Weiss theories are ad hoc, it was shown by Kac and collaborators in the 1950s[11] that the mean-field results could be derived rigorously if potentials of interaction of the right form were used. We will briefly describe the Kac approach in the language of magnetic systems, specifically Ising models. The Ising model is characterized by a lattice in d dimensions (e.g., a square lattice in $d = 2$ or cubic in $d = 3$) with a spin at each vertex. The Hamiltonian H is given by

$$-H = \sum_{ij} J_{ij}S_iS_j + h\sum_i S_i \tag{8}$$

where h is a uniform applied magnetic field, J_{ij} is the strength of the interaction between spins at sites i and j and S_i gives the configuration of the spin at the ith

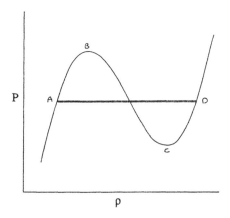

P

FIGURE 1 Schematic representation of the van der Waals equation for temperature T below the critical temperature T_C. The horizontal axis is the density and the vertical axis is the pressure. The Maxwell construction is the horizontal broken line.

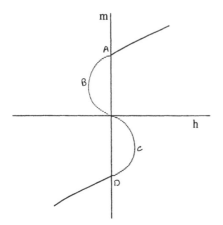

FIGURE 2   Schematic representation of the magnetization per spin m versus the applied magnetic field h for $T < T_C$ As in Figure 1, B and C are the spinodal points The vertical line A-D is the analog of the Maxwell construction and the metastable state.

site. For the Ising model $S_i = \pm 1$. If $J_{ij} > 0$ for all i and j, then the Ising model is ferro-magnetic and the spins will tend to align at low T. The partition function Z is given by

$$Z = \sum_{\text{spin configs.}} \exp[-\beta H] \tag{9}$$

where $\beta = 1/K_B T$ and K_B is Boltzmann's constant. The free energy in the canonical ensemble $F = -\beta^{-1} \ln(Z)$.

We will chose J_{ij} to be of the form

$$J_{ij} = \gamma^d \phi(\gamma x_{ij}) \tag{10}$$

where γ is a parameter and x_{ij} is the distance between spins at sites i and j. We take $\phi(\gamma x_{ij})$ to be of the form

$$\phi(\gamma x_{ij}) = 1, \ \gamma x_{ij} \le 1, \tag{11a}$$
$$\phi(\gamma x_{ij}) = 0, \ \gamma x_{ij} > 1. \tag{11b}$$

Kac and collaborators showed that if the partition function and free energy were calculated for a system with the Hamiltonian specified by Eqs. (8), (10), and (11) and the limit $\gamma \to 0$ taken that the magnetization $m = \partial F/\partial h$ had the Curie-Weiss form.

For the Kac form of the potential the metastable state is an analytic continuation of the stable state and the spinodal marks the limit of stability of the metastable state. The question of relevance to the slider-block models is: What happens to the spinodal if the Hamiltonian in Eq. (8) is used with the potential in Eq. (10) but the limit $\gamma \to 0$ is not taken? That is, we take $\gamma \ll 1$ but not equal to zero. The answer can be seen in Figure 3 where we have plotted the inverse of the isothermal susceptibility[9] measured in a $d = 3$ Ising model for various interaction ranges. The solid line is the theoretical prediction in the $\gamma \to 0$ limit and the different geometrical shapes denote data for various interaction ranges and hence q, the number of spins that interact with a single spin. The important point is that as the interaction range increases the system behaves more like the mean-field limit. A second important point is for $q = 342$, where the data is a very good approximation to the mean-field theory, the interaction range ($\gamma^{-1} \sim 6$) lattice spacings, which is not that large.

Another way of understanding the same point comes from the study of transfer matricies for Ising models with long-range interactions. Singularities, such as spinodals, appear in this formalism as intersection of the eigenvalues of the matrix when they are plotted as a function of the magnetic field h for fixed temperature T.[17] For mean-field models there is an eigenvalue crossing representing the spinodal at a real value of h which depends on T. For systems with finite-range interactions the

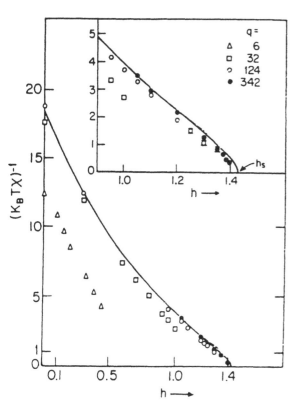

FIGURE 3 Inverse isothermal susceptibility χ (multiplied by Boltzmann's constant K_B and the temperature T) as a function of applied magnetic field h for different interactionranges for a $d + 3$ simple cubic Ising model. The parameter q. The solid line is the result of the meanfield calculation. The spinodal is at $h = 1.42$.

eigenvalue crossing occurs at complex values of both T and h. These results imply that the spinodal singularity is in the complex plane of the control parameters (T and h in the Ising model) for finite-range interaction but moves closer to the real axis as γ decreases finally reaching the real axis as $\gamma \to 0$. Consequently, for systems with long, but finite range, interactions the spinodal singularity is close enough to the real axis to affect the data as in Figure 3. We expect the system to become more mean-field like as γ decreases.

An aspect of spinodals that is important for understanding the results we present here is that as the temperature $T \to 0$ the spinodal approaches the co-existence curve. To understand why, one simply notes that if the system at $T \sim 0$ is aligned with all spins down, for example, then a spin will flip to the up state only if the applied positive field reaches a critical value. That value is determined by equating the energy of interaction of a spin with its neighbors with the interaction energy of the spin with the external field. Since $T \sim 0$, if one spin flips, they all will. Hence if we start at magnetic field $h = 0$ with all spins down and increase the field, either no spin will flip or they all will. The instability point is therefore at magnetization per spin $m = 1$. The important point is for $T = 0$ the strength of the net interaction of a spin with its neighbors sets the critical value of the field.

Before discussing our data, there is one additional concept that should be addressed; the relationship between thermal phase transitions (such as critical points and spinodals) and clustering or percolation. It has been a goal in Statistical Mechanics to express thermal quantities, such as free energies or correlation functions, in terms of clusters. The advantage of this approach is that a geometrical picture can facilitate intuitive understanding and makes possible the use of other tools such as fractals. Approximately 20–25 years ago several investigators tried to map thermal critical points onto a percolation transition.[16] For example, the nearest-neighbor interaction Ising model critical point was thought to be isomorphic to a correlated site percolation transition. This percolation transition occurs in a model where the spin at a site can take on two values $S_i = \pm 1$ and $S_i = +1$ is associated with an occupied site and $S_i = -1$ with an empty site. Two nearest-neighbor occupied sites are said to be connected, and a set of s mutually connected sites is called an s site cluster.

It was known in $d = 2$ that the Ising critical point corresponded to a percolation transition in this model but the critical exponents describing the thermal critical point differed from those of the percolation transition. In $d = 3$ the situation was worse in that the critical temperature T_c did not coincide with the percolation transition which occurs at $\sim .95T_c$.[16]

The problem was solved in 1980 when it was realized[5] that the proper percolation model had the additional feature of a random bond, with probability $p_b = 1 - \exp(-J/K_B T)$, between occupied sites. Two occupied sites are now connected only if there is an occupied bond between them. This correlated site—random bond model not only has the property that the percolation transition coincides with the critical point but the critical exponents that describe the clustering at the transition are identical to the thermal critical exponents. Consequently, it is the correlated site—random bond clusters that are the geometrical realization of the Ising thermal fluctuations.

A similar isomorphism can be made at the spinodal with the change that $p_b = 1 - \exp(-J\rho/K_B T)$ where ρ is the density of occupied sites.[14] The main message of this discussion is that clusters only reflect thermal properties of the system if they are properly chosen. We will return to these points in Section 5.

4. NUMERICAL RESULTS

In this section we present the results of our simulation of the slider-block cellular automata model presented in Section 2. We performed the simulations on the Boston University Connection Machine-5 using a configuration of either 16 or 32 RISC processors. The CM Fortran code updated all dynamical variables, such as displacement and stress on all blocks, in parallel. We have investigated systems of size 128^2 and 256^2 blocks arranged in an L^2 lattice with periodic, closed, open, and fixed boundaries. Most of the work that we have presented involves closed boundaries.

Periodic boundaries map the lattice onto a torus. A **closed** boundary block connects only to blocks within the lattice. An **open** or a **fixed** boundary block connects to "ghost" blocks outside of the lattice. For open boundaries, ghost blocks act as sinks for stress. For fixed boundaries, ghost blocks transfer ζ^F to the boundaries with each loader plate update, and then they act as sinks for stress as the system relaxes.

As we mentioned in Section 2, the long-range nature of the elastic Green's function implies a mean-field limit. We cannot of course simulate a system with $q = \infty$, so we present data with varying interaction and note the trends as the range is increased. Rather than the particular form of interaction in Eq. (11) we connect each block to blocks in a neighborhood defined by a square of side $2R + 1$, containing $q = (2R + 1)^2 - 1$ neighbors. The range of interaction R is proportional to γ^{-1}.

The system evolves as follows. Initially, the code assigns to each block a uniformly distributed random displacement with zero mean, and it also assigns the stress threshold to be either a constant distribution, $\sigma^F = 50$, or a random distribution with $\overline{\sigma_i^F} = 50$ and a predetermined variance. Immediately after the initialization, the loader plate moves, increasing the stress on each block by $K_L V$. The quantity Δ is taken equal to one. Blocks i with $\zeta_i \geq \zeta_i^F$ slide and change their positions according to the jump rule of Eq. (1) with $\zeta^R = 0$ for all i. A sliding block transfers qK_C/K of its stress to its neighbors, while $1 - qK_C/K$ of its stress leaves the system. Thus, stress is not conserved. This step is repeated until $\zeta_i < \zeta_i^F$ for all blocks i. Time is incremented by one unit. The loader plate increases stress again by $K_L V$ and the above process repeats.

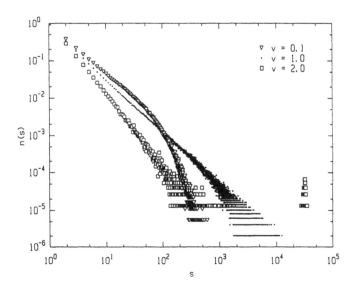

FIGURE 4   A log-log plot of the number of clusters $n(s)$ with s failed sites vs. s for $q = 24$. The failure threshold is 50 ± 10 Three values of V are plotted $V = 0.1, 1.0, 2.0$. The value $V = 1.0$ appears to be close to the critical velocity.

In Figure 4 we plot the log of the number of clusters with s blocks, $n(s)$ versus the log of the cluster size s for a system with $R = 2, q = 24$. Clusters are defined as follows: Assume that after a plate update there is exactly one block (which we will call an initiator) with $\zeta_i \geq \zeta_i^F$. After this block is moved, as discussed in Section 2, its neighbors (i.e., those attached to the ith block by a spring) may now have $\zeta_j \geq \zeta_j^F$. If so they are moved and are considered to be connected to the initiator and each other. The movement of these blocks may cause other blocks to have a stress greater than the failure threshold. These will be moved in turn and will also be considered connected to all blocks that have previously failed. If there is more than one initiator, that is more than one block with a stress greater than the failure threshold just after the plate update, several clusters will be generated simultaneously. Clusters will be said to coalesce and be counted as one if they share a failed block.

Figure 4 contains data for three different velocities of the loader plate, $V = 0.1$(triangles), 1.0(small dots), and 2.0(squares). For $V = 0.1$ $n(s)$ vs. s exhibits behavior typical of systems near critical points. In particular there is a part of the curve, from $s = 1$ up to about $s = 100$ that can be fit to a straight line but the larger s range exhibits an exponential cutoff.

For $V = 1.0$ the curve has straightened out substantially and can now be fit to a straight line up to $s \sim 10^3$. For $V = 2.0$ the portion of the curve that can be fit to a straight line has decreased. In addition clusters the size of the system have appeared indicating that some percolation threshold has been passed. Our interpretation of this data is that there is a percolation[16] transition in the neighborhood of $V = 1.0$. The slope of the straight line fit to the data for $V = 1.0$ is ~ -1.7 which is close to the cluster scaling exponent for mean-field percolation.[16]

In Figure 5 we present the cluster scaling data for the same system but with $R = 4$. The same general trend is seen except that it is shifted to lower values of V. For $V = 0.01$ we see a pronounced exponential cutoff, the scaling plot straightens considerably for $V = 0.3086$ and for $V = 1.0$ the cutoff again goes to smaller values of cluster size s with the addition of "infinite" clusters with a diameter the size of the system. This behavior is the same as we saw for $R = 2$ except that the "critical" velocity is lower.

The question one would naturally ask once this scaling or power-law form is established is whether the scaling is associated with some underlying critical phenomena. This is an important question because, if there is an underlying critical point that is responsible for the observed scaling, then there are large spatial and temporal correlations which might account for the spatiotemporal clustering in earthquakes.

We are interested in, for example, the fluctuations in the mean slip $\bar{U}$ which is defined as

$$\bar{U} = \frac{1}{T} \int_0^\infty dt \frac{1}{N} \sum_i [U_i(t) - \bar{U}_i]^2 \tag{12}$$

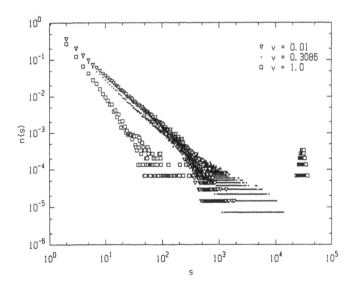

FIGURE 5 Same plot as in Figure 4 for $R = 4$, $q = 80$. Three values of V are plotted; $V = 0.01, 0.3086, 1.0$. The value $V = 0.3086$ appears to be close to the critical velocity.

where

$$\bar{U}_i = \frac{1}{T} \int_0^\infty dt U_i(t) \tag{13}$$

and N is the number of blocks in the system. If the cluster scaling of Figures 4 and 5 is associated with a critical point then we would expect to see large fluctuations in the amount that the center of mass of the blocks lags behind the loader plate. This would be reflected in the divergence of $\bar{U}$ as the critical point is approached. In order to investigate this point as well as the existence of divergent time scales associated with the critical point,[19] we measure the power spectrum associated with the slip.

The power spectrum $P(\omega)$ associated with the slip is defined through the temporal Fourier transform, $\Delta U(\omega)$, of $\Delta U(t) = \sum_i [U_i(t) - \bar{U}_i]$

$$P(\omega) = \langle \Delta U(\omega) \Delta U(-\omega) \rangle \tag{14}$$

where the brackets $\langle \rangle$ denote the ensemble average. Since the power spectrum is the zero-momentum transfer limit of the wave number and frequency-dependent structure factor, $S(k, \omega)$, the $\omega \to 0$ limit is related in equilibrium via the fluctuation dissipation theorem to the susceptibility which indicates the size of the fluctuations in the slip.[7]

In addition to the susceptibility we can also obtain a characteristic relaxation time from the power spectrum. Since the power spectrum is the Fourier transform of the slip autocorrelation function,[18] it is expected to be of the form

$$P(\omega) = \chi \frac{\tau^{-1}}{\omega^2 + \tau^{-2}} \tag{15}$$

where τ is a characteristic relaxation time and χ is the measure of the fluctuations in the slip deficit.

In Figure 6 we plot $\ln P(\omega)$ vs. $\ln \omega$ for $R = 2$. If the sequence of slips is associated with a critical point, then $\tau \gg 1$ and we expect this plot to be a straight line characteristic of power law. At $\omega = 0, P(\omega)$ will saturate at τ. As can be seen from Figure 6(a) for $V = 0.01$ the power spectrum shows some signs of a straight line characteristic of a critical point. However, at this velocity the cluster plots in Figure 4 show that the cluster scaling has an exponential cutoff at a rather small cluster size. In Figure 6(b) we show the power spectrum for $V = 1.0$, the velocity at which the cluster plot exhibits behavior closest to what we expect at a critical point. As can be seen from the figure the power spectrum is flat indicating a power-law decay in the autocorrelation function with a small decay time. In Figure 6(c) we plot the power spectrum for $V = 2.0$, well above critical scaling in the cluster plots, which is also flat.

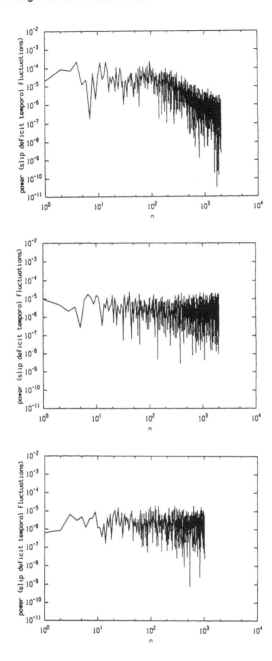

FIGURE 6   The power spectrum associated with the slip deficit for $R = 2$, $q = 24$.
(a) $V = 0.1$, (b) $V = 1.0$, (c) $V = 2.0$. The velocity $V = 0.1$ appears to be closest to a
critical point spectrum.

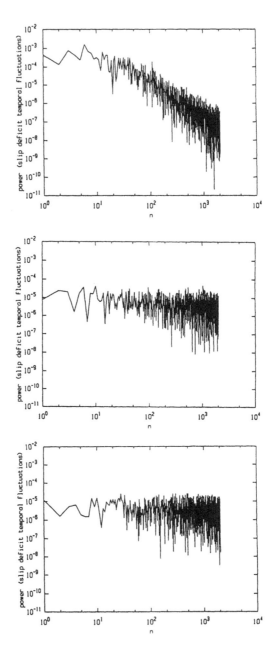

FIGURE 7 The power spectrum for $R = 4$, $q = 80$. (a) $V = 0.01$, (b) $V = 0.3086$ (c) $V = 1.0$. The spectrum for $V = 0.01$ appears to be closest to critical.

In Figures 7(a)–(c) we present the corresponding power spectra for $R = 4$. Note that the same pattern emerges as in the $R = 2$ case of Figure 6 but that the time scale for which the spectrum looks critical for $R = 4$, $V = 0.01$ is considerably longer than the corresponding $V = 0.1$ for $R = 2$.

How can we interpret the result that the power spectrum is flat at the value of the velocity associated with power-law cluster scaling? Certainly at this velocity for either $R = 2$ or $R = 4$ there is no thermal critical point. In fact the thermal critical point, or spinodal, appears to be at $V = 0$ in both cases. If that is true, then for any $V > 0$ the system should be unstable. In order to check this point we measured a quantity called the energy metric.[26] The metric measures the difference between the time average of a quantity, in this case the energy of each block in the system, and the ensemble average, the time averaged energy of each block averaged over the system. Mathematically this metric $\Omega_e(t)$ is given by

$$\Omega_e(t) = \frac{1}{N} \sum_{i=1}^{N} \left[\epsilon_i(t) - \bar{\epsilon}(t) \right]^2 \tag{16}$$

where

$$\epsilon_i(t) = \frac{1}{t} \int_0^t \epsilon_i(t')dt' \tag{17}$$

and

$$\bar{\epsilon}(t) = \frac{1}{N} \sum_i^N \epsilon_i(t). \tag{18}$$

If the system is ergodic $\Omega_e(t) \sim 1/t$.[26] In Figures 8(a)–(c) we plot the energy metric associated with the Rundle-Brown cellular automata model for $R = 4$ for several velocities. As can be seen from the figures, the system is not ergodic at any velocity. This is consistent with the idea that $V = 0$ is the spinodal.

Another check on this idea is the relationship between the power spectrum and the zero-momentum transfer limit of the structure factor. The equality of these two quantities relies on the system being in equilibrium. In Figure 9 we plot a scaled version of the power spectrum for $V = 0.0025, 0.005$ and 0.01. For $V = 0.01$ we have not scaled the data but for $V = 0.005$ and 0.0025 we have sampled every second and fourth time step respectively. As can be seen from the figure, the curves lie on top of each other.

Two implications can be drawn from this data. The first is that the time scale associated with the straight line behavior of $P(\omega)$ diverges as V^{-1}. The second has to do with the fact that the value of $P(\omega)$ in the $\omega \to 0$ limit does not increase as expected as a critical point is approached **if the system is in equilibrium**. We conclude that the system is not in equilibrium consistent with the measurements of the metric in Figure 8.

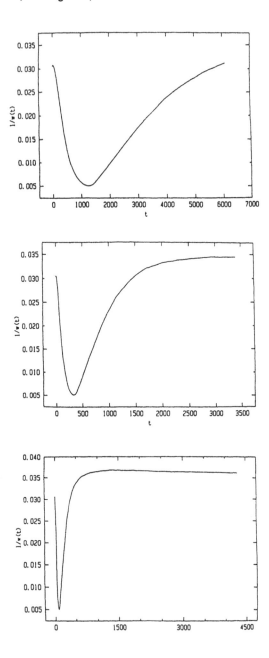

FIGURE 8   The inverse of the energy metric vs time for $R = 4$, $q = 80$. (a) $V = 0.0$ (b) $V = 0.075$, (c) $V = 0.3086$. None of the figures can be fit to a straight line over any time interval.

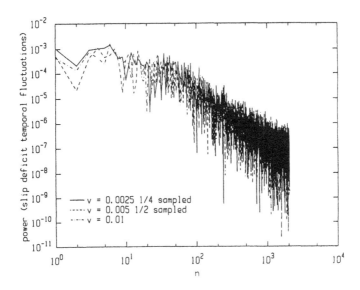

FIGURE 9   Scaled power spectrum for $R = 4$, $q = 80$. For $V = 0.01$ all points (plate updates) are used in the power spectrum. For $V = 0.005$ every other point is used and the total number of plate updates is doubled so that the frequencies are the same for the same value of n. For $V = 0.0025$ every fourth point is sampled and the number of plate updates is scaled accordingly.

5. PHYSICAL CONSIDERATIONS

In this section we will outline some of the physical reasons why we interpret the cluster and power spectrum data as indicating a spinodal at $V = 0$. We stress that we are only considering the system with no noise and that the presence of noise will change some of our conclusions. We will discuss the model with noise very briefly in the next section.

In order to understand why the percolation threshold in this model should approach zero as $R \to \infty$, we first note that after the system has evolved enough so that the transient effects associated with the initial random distribution of the blocks have decayed the distribution of the stress relative to the failure threshold is flat. In particular, the average number of blocks brought to failure by each plate update is constant.

When a block fails, it transmits ζ^F/q units of stress to its neighbors (to order $1/q$) if the threshold is flat and $\bar{\zeta}^F/q$ if the threshold is random. Here $\bar{\zeta}^F$ is the average threshold. Since the fraction of blocks in a stress interval is $1/\zeta^F$, then the density of failed blocks is q^{-1}. The probability then that there is a block that will fail within the "interaction range" of a failed block is $q/q = 1$. This implies that there is a percolation path to infinity for any V hence the critical velocity for percolation is zero in the mean-field limit consistent with our data. The argument is mean field since we ignored terms of order $1/q^2$ and higher and also ignored fluctuations.

In order to see why there should be a "thermodynamic" spinodal at $V = 0$ we note that since there is no noise any block brought to failure by moving the plate will cause a runaway failure event. In fact since the amount of stress put into the system due to a plate update is given by NV where N is the number of blocks and $K_L = 1$ each block will have to fail q times to relieve the stress. This is because the loader plate springs are the only way stress can enter or leave the system. The springs that connnect blocks merely redistribute stress when a block fails. The amount of stress dissipated by the loader plate spring is $1/K_T \sim 1/(1 + q)$ which goes to zero in the mean-field limit. Consequently, for any finite velocity blocks will fail q times which becomes infinite in mean field.

6. SUMMARY AND CONCLUSIONS

From the data presented in the last section we conclude that the cellular automata version of the Burridge-Knopoff model has a spinodal singularity associated with the cluster scaling in the mean-field limit. In the zero noise case we considered, the spinodal as well as the percolation threshold will be at $V = 0$.

The mean-field limit arises because the stress Green's function, whose form is taken from linear elasticity theory, falls off as r^{-3}. This long-range "interaction" imposes the mean-field behavior.

The fact that the critical point is actually a spinodal singularity has implications for the nature of earthquake events and their statistical distribution and leads to interesting conclusions about the way faults self organize.

In order to understand the implications of the existence of the spinodal for these models it is important to realize that the spinodal is associated with the limit of stability of a metastable state. For the case of no noise the spinodal is at $V = 0$, but as noise is added the spinodal velocity should rise as the value of the spinodal magnetic field rises with rising temperature in spin models.[13,15] This suggests that the system with noise is sitting in a metastable well near a spinodal instability if V is not too large. If R is big enough, the metastable state will last a long time even if the well is shallow. The reason for this behavior is that the barrier to nucleation events, which drive the system out of the metastable state, is proportional to R^d [13,27] where d, the dimension of space is 2 in the case of planar faults. The system will, for the lifetime of the metastable state, be ergodic and the number of events will scale as in the Gutenberg-Richter plot. Moreover, the distribution of events will be given by a Boltzmann factor[24] (also see article by Rundle et al. in this volume). As this system continues to evolve, a nucleation event will eventually occur. This very large event will drive the system out of the quasi-equilbrium state causing a breakdown in ergodicity. Since nucleation events have a characteristic waiting time particularly in systems with long-range forces,[13,27] these events will be quasi-periodic. If the velocity is such that the spinodal, in a system with noise, is approached too closely, space-time clustering caused by the critical point will be prominent.

Studies of this system with random noise have begun,[24] and we will continue to probe the effects of noise, velocity of plate and roughness on the statistical distribution of "earthquake" events.

In summary we have presented data and arguments that support the existence of a spinodal singularity in cellular automata models of earthquake faults. The spinodal is associated with a cluster scaling similar to that of Gutenberg-Richter. Contrary to short-range models, the model that is perhaps most realistic, the one with long-range forces, has the spinodal at zero velocity for no noise and most probably at a very low velocity for noise levels that might exist on real faults. The existence of the spinodal and its influence on the statistical distribution of "earthquake events" might explain the variety of behavior seen on real faults.

ACKNOWLEDGMENTS

The work of WK and CF was supported by Department of Energy grant DE-FG02-95ER14498. The work of JBR was supported by Department of Energy grant DE-FG03-95ER14499. The authors would like to acknowledge the assistance of the Boston University Center for Computational Science and the Center for Information Technology. WK and JBR would like to acknowledge the hospitality of the Santa Fe Institute where much of this work was done.

REFERENCES

1. Bak, P., and C. Tang. *J. Geophys. Res.* **94** (1989): 15635.
2. Burridge, R., and L. Knopoff. *Bull. Seism. Soc. Am.* **57** (1967): 341.
3. Carlson, J. M., and J. S. Langer. *Phys. Rev.* **A 40** (1989): 6470.
4. Carlson, J. M., and J. S. Langer. *Phys. Rev.* **A 44** (1991): 884.
5. Coniglio, A., and W. Klein. *J. Phys.* **A 13** (1980): 2775.
6. Feder, H. J. S., and J. Feder. *Phys. Rev. Lett.* **66** (1991): 2669.
7. Forster, D. In *Hydrodynamic Fluctuations, Broken Symmetry and Correlation Functions.* Reading, MA: Benjamin, 1975.
8. Gu, J. C., J. R. Rice, A. L. Ruina, and S. T. Tse. *J. Mech. Phys. Solids* **32** (1984): 167.

9. Heermann, D. W., W. Klein, and D. Stauffer. *Phys. Rev. Lett.* **49** (1982): 1262.
10. Huang, J., and D. L. Turcotte. *Nature* **348** (1990): 234.
11. Kac, M., G. E. Uhlenbeck, and P. C. Hemmer. *J. Math. Phys.* **4** (1961): 216.
12. Kanamori, H., and D. L. Anderson. *Bull. Seism. Soc. Am.* **65** (1975): 1073.
13. Klein, W., and C. Unger. *Phys. Rev.* **B 28** (1983): 445.
14. Klein, W. *Phys. Rev. Lett.* **65** (1990): 1462.
15. Lebowitz, J. L., and O. Penrose. *J. Math. Phys.* **7** (1966): 98.
16. Novotny, M., W. Klein, and P. Rikvold. In *Introduction to Percolation Theory*, edited by D. Stauffer et al. London and Philadelphia, 1985.
17. Novotny, M., W. Klein, and P. Rikvold. *Phys. Rev.* **B 33** (1986): 7729.
18. Robertson, H. S. In *Statistical Thermophysics*. Englewood Cliffs, NJ: Prentice Hall, 1993.
19. Rundle, J., and W. Klein. In *Introduction to Phase Transitions and Critical Phenomena*, edited by H. E. Stanley. New York: Oxford University Press, 1971.
20. Rundle, J. B., and D. D. Jackson. *Bull. Seism. Soc. Am.* **67** (1977): 1363.
21. Rundle, J. B., and S. R. Brown. *J. Stat. Phys.* **65** (1991): 403.
22. Rundle, J., and W. Klein. In *Proceedings of the 33rd Symposium on Rock Mechanics*, edited by J. R. Tillerson and W. R. Wawwesik. Protterdam: A. A. Balema, 1992.
23. Rundle, J. B., and W. Klein. *J. Stat. Phys.* **72** (1993): 405.
24. Rundle, J. B., W. Klein, S. Gross, and D. Turcotte. *Phys. Rev. Lett.* **75** (1995): 1658–1661.
25. Scholz, C. H., ed. *The Mechanics of Earthquakes and Faulting*. Cambridge, MA: Cambridge University Press, 1990.
26. Thirumalai, D., and R. Mountain. *Phys. Rev.* **E 47** (1993): 479.
27. Unger, C., and W. Klein. *Phys. Rev.* **B 29** (1984): 2698.

W. I. Newman[1], D. L. Turcotte[2], and A. Gabrielov[3]
[1]Departments of Earth and Space Sciences, Physics and Astronomy, and Mathematics, University of California, Los Angeles, CA 90095-1567
[2]Department of Geological Sciences, Snee Hall, Cornell University, Ithaca, NY 14853
[3]Mathematical Sciences Institute, Cornell University, Ithaca, NY 14853

A Hierarchical Model for Precursory Seismic Activation

Seismic activation has been recognized to occur before many major earthquakes. Seismic activation occurred in the San Francisco Bay area, particularly in the East Bay, prior to the 1906 earthquake. There is a serious concern that the recent series of earthquakes in southern California is seismic activation prior to a great southern California earthquake. Seismic activation in a broad region forms a primary basis for the intermediate-term earthquake prediction carried out by the Moscow group under Academician V. I. Keilis-Borok. Bufe and Varnes[5] quantified seismic activation prior to the Loma Prieta earthquake in terms of a power-law increase in the regional Benioff strain release prior to this earthquake. Sornette and Sammis[23] have considered this data and concluded that there is an excellent fit to a log-periodic increase in the Benioff strain release. In order to better understand activation a hierarchical seismic failure model has been studied. An array of stress-carrying elements is considered (formally, a cellular automaton or lattice gas, but analogous to the strands of an ideal, frictionless cable). Each element has a time-to-failure that is dependent on the stress the element carries and has a statistical distribution of values.[8,9]

Reduction & Predictability of Natural Disasters, Eds. Rundle, Turcotte, & Klein,
SFI Studies in the Sciences of Complexity, Vol. XXV, Addison-Wesley, 1996   **243**

When an element fails, the stress on the element is transferred to a neighboring element; if two adjacent elements fail, stress is transferred to two neighboring elements; if four elements fail, stress is transferred to four adjacent elements, and so forth. When stress is transferred to an element its time to failure is reduced. The stress redistribution technique models the transfer of stress from faults of varying sizes. The reduced times to failure are consistent with the aftershock decay following an earthquake (Omori's law). The intermediate-size failure events prior to total failure each have a sequence of precursory failures, and these precursory failures each have an embedded precursory sequence of smaller failures. The entire failure sequence is "fractal," each failure appears to have a self-similar sequence of failures on a smaller scale. The total failure of the array appears to be a critical point. One observation is that there is a sequence of partial failures leading up to the total failure that resembles a log-periodic sequence.

INTRODUCTION

Our present understanding of earthquakes is, at best, limited. There are two fundamental related problems which form the basis of the geomechanics of earthquakes and seismicity. The first of these concerns the mechanics of the faults on which earthquakes occur. How does the earthquake rupture initiate and how does it propagate? The second question concerns the interactions between faults. The frequency-size distribution of earthquakes in a region is self-similar or "fractal" and earthquakes are accepted to be a type of self-organization.

We first discuss the initiation and propagation of a rupture on a fault. The concept of a static coefficient of friction is generally applied to the initiation of rupture. Once the rupture initiates a dynamic coefficient of friction is applied. Many solutions for this problem have been obtained assuming a single fault in a homogeneous stress field, i.e., Tse and Rice.[25] However, there are a number of serious problems with this approach. One major problem is the stress level. Observations (i.e., Zoback and Healy[28]) favor stress levels associated with failure on faults that are nearly an order of magnitude lower than the values predicted by laboratory friction experiments. This and other problems led Smalley et al.[22] to propose a hierarchical failure model for the initiation and propagation of failure on a fault. The asperities on a fault were treated as individual elements with a probabilistic distribution of strengths. If one element failed, the stress was transferred to the adjacent element on which an induced failure could occur. If two elements failed, the stresses were transferred to two adjacent elements and so forth. A cascade of failures occurred.

PRECURSORY ACTIVATION

A universal feature of regional seismicity is that it satisfies the Gutenberg-Richter frequency-magnitude relation

$$\log_{10} N = a - bM_L \tag{1}$$

with N the cumulative number of earthquakes with magnitude greater than M_L and a and b constants. The dependence of N on M_L for southern California for each year between 1980 and 1994 is given in Figure 1. In general there is good agreement with Eq. (1) taking $a = 4.3$ and $b = 1.06$. The exceptions can be attributed to the aftershock sequences of the Whittier (1987), Landers (1992), and Northridge (1994) earthquakes.

The near uniformity of the background seismicity in southern California is clearly striking. This is strongly suggestive of thermodynamic behavior and self-organization. A near universal feature of critical phenomena is action at a distance. There is increasing evidence that such long-distance correlations are a characteristic of seismicity and are related to seismic activation.

Although long-distance correlations between earthquakes have generally been rejected in the United States and Japan, such correlations have been widely accepted in China and Russia as well as the former Soviet Union. A striking example was a sequence of five earthquakes that occurred in China between 1966 and 1976. These were the $m = 7.2$ Shentai (1966), $m = 6.3$ Hijien (1967), $m = 7.4$ Bo Sea (1969), $m = 7.3$ Haicheng (1975), and the $m = 7.8$ Tangshan (1976). These earthquakes spanned a distance of some 700 km and the Haicheng earthquake was successfully predicted by the Chinese, at least partially on the basis of seismic activation.[20]

Seismic activation has been previously recognized in association with an increase in seismicity that occurred in the San Francisco Bay area prior to the 1906 earthquake.[24] Earthquakes with estimated magnitudes between 6.5 and 7.0 occurred in 1865 (Santa Cruz Mountains), 1868 (Hayward), 1892 (Vacaville), and 1898 (Mare Island). There is a serious concern that a similar seismic activation is now underway in southern California. A number of intermediate-size earthquakes have occurred in southern California in the last 45 years. These include the magnitude 7.4 Kern County earthquake on July 21, 1952, the magnitude 6.4 San Fernando earthquake on February 9, 1971, the magnitude 7.6 Landers earthquake on June 28, 1992, and the magnitude 6.6 Northridge earthquake on January 17, 1994.

Long-distance correlations and seismic activation form the basis of the pattern recognition earthquake prediction algorithms developed by a group of earthquake probabilists and statisticians in Moscow working under the direction of

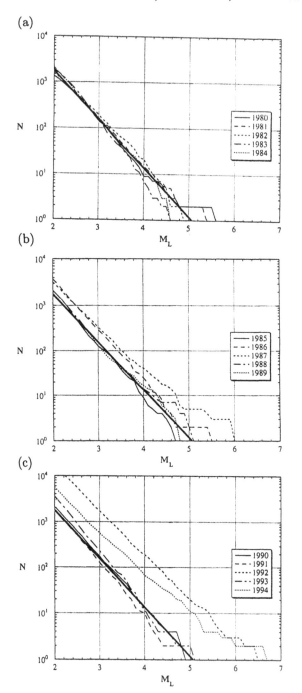

FIGURE 1   The cumulative number of earthquakes N with magnitude greater than M_L for each year between 1980 and 1994 is given as a function of M_L, the region considered is southern California. The straight line correlation is the Gutenberg-Richter relation with $a = 4.3$ and $b = 1.06$.

Vladimir Keilis-Borok. The pattern recognition included increases in regional seismicity, increases in the clustering of earthquakes, and changes in aftershock statistics. Premonitory seismicity patterns were found for strong earthquakes in California and Nevada (algorithm "CN") and for earthquakes with $M > 8$ worldwide (algorithm "M8"). Assuming self-similarity of the properties of seismicity, both algorithms were tested in seismically active regions.[13] When observed levels exceed pre-established levels, intermediate-term (1–3 year) predictions were made. TIP's (times of increased probability) were formally issued. During the last eight years, eight strong earthquakes (including Loma Prieta, Landers, and Northridge) were predicted in advance.

This approach is certainly not without its critics. Independent studies have established the validity of the TIP for the Loma Prieta earthquake; however, the occurrence of recognizable precursory patterns prior to the Landers earthquake are questionable. Also, the statistical significance of the size and time intervals of warnings in active seismic areas has been questioned. Nevertheless, seismic activation prior to a major earthquake certainly appears to be one of the most promising approaches to earthquake prediction.

Although long-distance correlations have generally been rejected in the United States, the Landers earthquake provided direct evidence that faults interact with each other over large distances.[10] The Landers earthquake triggered earthquakes at 14 distant sites scattered over the western United States. The furthest site was Yellowstone National Park in Wyoming, 1250 km from Landers. The largest triggered earthquake, magnitude 5.6, occurred 250 km from Landers. Such long distance correlations are a primary characteristic of critical phenomena.

Another approach to earthquake prediction based on long distance correlations has been given by Sornette and Sammis.[23] This is an extension of an approach to earthquake prediction given by Varnes,[27] Bufe and Varnes,[5] and Bufe et al.[4] These authors suggested that there was a power-law increase in the regional cumulative Benioff strain release prior to an earthquake. The data given by Bufe and Varnes[5] for the Loma Prieta earthquake is shown in Figure 2. Sornette and Sammis[23] considered this data and concluded that there is an excellent fit to a log-periodic increase in seismic activity.

A simple power-law (fractal) increase in the cumulative Benioff strain E is given by

$$E = C(t_f - t)^\alpha, \tag{2}$$

where $t_f - t$ is the time prior to the earthquake and the constant α is negative. Bufe and Varnes[5] used this relation to predict the time t_f when the earthquake would occur. Consider the case when the exponent is complex: $\alpha = \xi + i\eta$. In this case Eq. (2) becomes

(a)

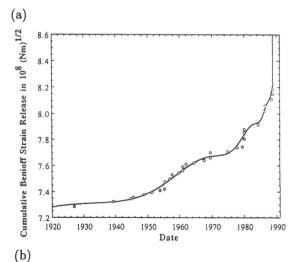

(b)

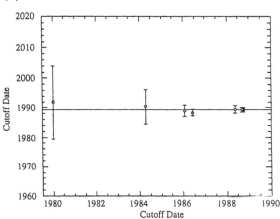

FIGURE 2 (a) The data points are the cumulative Benioff strain release in magnitude 5 and greater earthquakes in the San Francisco Bay area prior to the October 17, 1989, Loma Prieta earthquake.[5] The solid line is the log-periodic correlation with the data. (b) Predicted dates for the Loma Prieta earthquake based on the data applicable prior to the cutoff date.[23]

$$E = \Re\left\{C\left(t_f - t\right)^{\xi + i\eta}\right\}$$
$$= \Re\left\{C\left(t_f - t\right)^{\xi} \exp\left[i\eta \ln\left(t_f - t\right)\right]\right\} \qquad (3)$$
$$= C\left(t_f - t\right)^{\xi} \cos\left[\eta \ln\left(t_f - t\right)\right]$$

where $\Re$ stands for the real part. This is log-periodic behavior. Results from condensed matter physics imply that additional terms with this increased complexity appear in the expression, supplementing the simple form in Eq. (2).

The sequence of positive maxima in Eq. (3) corresponds to the sequence

$$t_f - t_n = \exp \left[\frac{1}{\eta} \tan^{-1} \left(\frac{\xi}{\eta} \right) + \frac{\pi n}{\eta} \right] , \tag{4}$$

where $n = 1, 2, 3, \ldots$ If three successive values of the maxima are observed, t_1, t_2, t_3, the failure time t_f is given by

$$t_f = \frac{t_2^2 - t_1 t_3}{2t_2 - t_1 - t_3} . \tag{5}$$

Thus, successive values of maxima can *in principle* be used directly to predict failure times.

Sornette and Sammis[23] introduced a more general form

$$E = A + B \left(t_f - t \right)^\xi \left\{ 1 + C \cos \left[\eta \ln \left(t_f - t \right) + \theta \right] \right\} , \tag{6}$$

in keeping with a result in perturbation theory where additional terms like those in Eq. (3) must appear. The correlation of this result with the data of Bufe and Varnes[5] is illustrated in Figure 2, and excellent agreement is found. This relation was used to obtain retrospective predictions for the occurrence of earthquakes and their predictions for the Loma Prieta earthquake are also illustrated in Figure 2.

The concept of self-organized criticality (SOC) was introduced by Bak et al.[2] SOC is defined to be a natural system in a marginally stable state that, when perturbed from that state, will evolve naturally back to the state of marginal stability. Energy input to the system is continuous but the energy loss is in a discrete set of events that satisfy self-similar or fractal frequency-size statistics. Distributed seismicity is taken to be the classic example of a natural system that exhibits self-organized criticality. There is a continuous input of energy (strain) through the relative motion of tectonic plates. This energy is dissipated in a fractal distribution of earthquakes, the Gutenberg-Richter relation Eq. (1) being equivalent to a fractal (power-law) relation between earthquake frequency and rupture area.[1] Scholz[21] has argued that the earth's entire crust is in a state of self-organized criticality. He makes the point that, wherever a large dam is built, induced seismicity results from the filling of the reservoir. Thus the crust is everywhere on the brink of failure. A variety of models that exhibit self-organized critical behavior (although the issue of "criticality" in these models remains somewhat controversial) yield similar statistics. One example is a two-dimensional array of slider blocks; the blocks are pulled over a surface by driver springs connected to a constant velocity driver plate, adjacent blocks are connected by connector springs. The blocks interact with the surface plate with a prescribed static-dynamic friction law. Many numerical studies of arrays have been carried out (i.e., Carlson and Langer[6]; Carlson et al.[7]; Nakanishi[15,16]; Brown et al.[3]; Ito and Matsuzaki[12]; Huang et al.[11]). Fractal (i.e., self-similar) frequency-size statistics of slip events are generally found. It is concluded that stress transfer between a hierarchical distribution of faults in the earth's crust is an essential feature of distributed seismicity.

In order to examine the questions discussed above we consider a hierarchical, time-to-failure model for precursory activation. A series of elements are considered with each element having a prescribed lifetime. When elements fail, the stress on the elements is transferred to adjacent elements. This class of models has been applied to fibre bundles and composite materials (Phoenix and Tierney[19]; Newman and Gabrielov[17]; Newman et al.[18]). We will show that this model incorporates many of the features of distributed seismicity and yields an activation prior to total failure that resembles a log-periodic sequence.

THE MODEL

Our hierarchical model for failure is illustrated in Figure 3. It is a one-dimensional analog model for failure due to stress transfer. At the lowest order in this example there are 128 zero-order elements. These elements are paired to give 64 first-order elements, the 64 first-order elements are paired to give 32 second-order elements and so forth. A statistical distribution of lifetimes is assigned to the lowest order elements. When one of these elements fails, the stress on the element is transferred to the neighboring element, increasing the stress on it. If a pair of zero-order elements fail, i.e., a first-order element, the stress is transferred to the adjacent pair of zero-order elements, i.e., to the adjacent first-order element, and so forth.

In order to illustrate the stress transfer consider the second-order ($n = 4$) example given in Figure 4. Each element is given a probabilistic "lifetime" and two examples of failure are illustrated. At time $t = 0$ the stress σ_0 is applied to the four elements. In both realizations element "2" has the shortest life time and it is the first to fail. The stress σ_0 on element "2" is transferred to element "1" placing a stress $2\sigma_0$ on this element as illustrated in Figure 4 (ii). The question now is whether the enhanced stress on element "1" will cause it to fail prior to elements "3" or "4." In realization (a) element "1" is the next to fail and the stress $2\sigma_0$ on this element is transferred to elements "3" and "4" placing a stress $2\sigma_0$ on both of these elements. Element "4" is next to fail and the stress $2\sigma_0$ on it is transferred to the last surviving element "1" which has a stress $4\sigma_0$. In realization (b) element "4" next to fail and the stress σ_0 on this element is transferred to element "3" placing a stress $2\sigma_0$ on this element. Element "3" is the next to fail and the stress $2\sigma_0$ is again transferred to the last surviving element "3" which has a stress $4\sigma_0$.

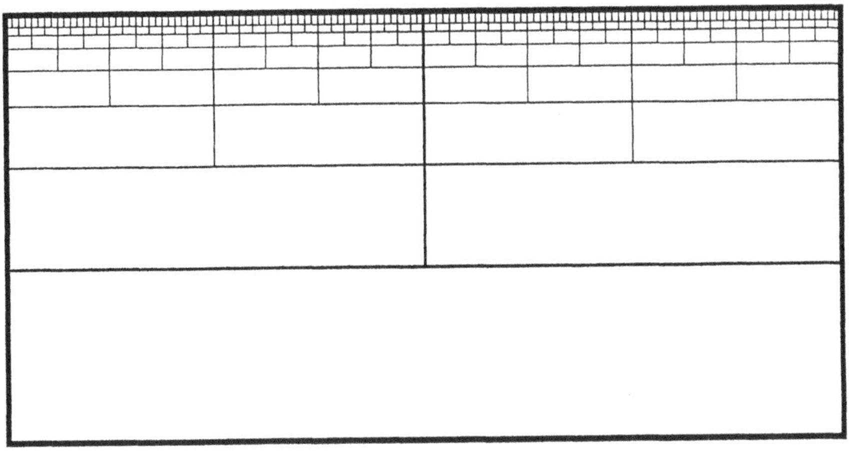

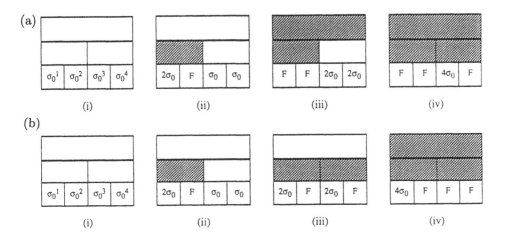

FIGURE 3 Illustration of a seventh-order ($N = 128$) example of our hierarchical model.

FIGURE 4 Illustration of stress transfer in a second order ($N = 4$) example of our hierarchical model. Each element is given a statistical "lifetime." In example (a) element "2" fails transferring stress to element "1," element "1" then fails and stress is transferred to elements "3" and "4," element "4" fails transferring stress to element 3 that subsequently fails. In example (b) element "2" fails transferring stress to element "1," element "4" fails transferring stress to element "3," element "3" fails transferring stress to element "1" that subsequently fails.

The zone of stress transfer is equal in size to the zone of failure. This local load-shearing model simulates the Green's function associated with the elastic redistribution of stress adjacent to a rupture.

Before formulating the local load-sharing model for failure we will illustrate the approach by considering a simple global load-sharing model. Initially we have N_0 elements each carrying a load σ_0. When an element fails, the load on that element is transferred uniformly to all the remaining elements. When n_f elements have failed the stress σ on the $n_s = N_0 - n_f$ surviving elements is given by

$$\sigma = \frac{N_0}{N_0 - n_f} \sigma_0 . \tag{7}$$

The rate at which elements fail is assumed to be given by the rate law

$$\frac{d(N_0 - n_f)}{dt} = -\nu(N_0 - n_f) \tag{8},$$

where the hazard rate ν is related to the stress by

$$\nu = \nu_0 \left(\frac{\sigma}{\sigma_0}\right)^\rho , \tag{9}$$

where ν_0 is the hazard rate of a single element under load σ_0 and the power ρ is typically in the range of 2–5. Combining Eqs.(7)–(9) gives

$$\frac{d(N_0 - n_f)}{dt} = -\nu_0 \frac{N_0^\rho}{(N_0 - n_f)^{\rho-1}} . \tag{10}$$

Integrating with the condition that $n_f = N_O$ when $t = t_f$ we obtain

$$n_f(t) = N_0 \left\{1 - [\rho\nu_0(t_f - t)]^{1/\rho}\right\} . \tag{11}$$

The number of surviving elements has a power-law dependence on the time to failure with the exponent ρ^{-1}. This result is clearly similar to the power-law relation for increase in Benioff strain given in Eq. (2).

We now determine failure statistics for the hierarchical model illustrated in Figure 3. Before obtaining numerical simulations it is necessary to prescribe the failure statistics for the individual elements. The failure of an engineering material is generally modeled in terms of a statistical distribution of lifetimes when subject to an applied stress σ_0 (Coleman[8,9]). The cumulative distribution of failure times t_f for an individual element can be written as

$$F(t_f) = 1 - \exp(-\nu_0 t_f) , \tag{12}$$

where $\nu_0 (\sigma_0)$ is the hazard rate under stress σ_0. This distribution of failure times is illustrated in Figure 5, the mean lifetime of an element is $t_{1/2} = \nu_0^{-1} \ln 2$. Each of the N elements is assigned a failure time t_{i0} based on Eq. (12) for an applied stress σ_0. The statistical representation given in Eq. (12) is entirely equivalent to the failure statistics obtained from the rate law Eq. (8). Using the stress dependence introduced in Eq. (9) a Weibull distribution for failures is obtained

$$F\left(t_f\right) = 1 - \exp\left[-\nu_0 \left(\frac{\sigma}{\sigma_0}\right)^\rho t_f\right] . \tag{13}$$

If elements are subjected to a constant stress σ at $t = 0$, Eq. (9) gives the statistical distribution of failure times t_f. The Weibull distribution is found to be in agreement with experiments on a wide variety of materials with ρ typically in the range 2–5.

However, with stress transfer the stress is not necessarily constant. In order to accommodate the increase in stress caused by local load sharing from failed elements we introduce a reduced time to failure for each element T_{if} for each element given by

$$t_{i0} = \int_0^{T_{if}} \left[\frac{\sigma\left(t\right)}{\sigma_0}\right]^\rho dt . \tag{14}$$

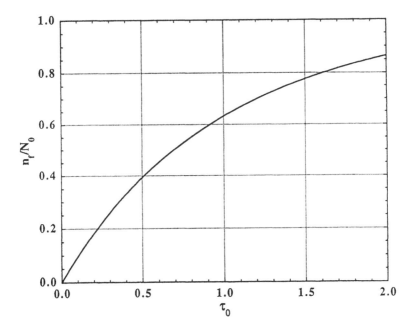

FIGURE 5 Illustration of the cumulative distribution of nondimensional failure times $\nu_0 t_0$.

(a)

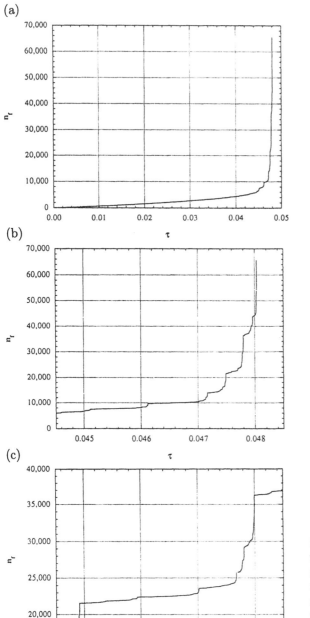

(b)

(c)

FIGURE 6 Failure sequence for a 16-order ($N = 65,536$) realization of our model. (a) Entire failure sequence (failure is completed at $\tau = 0.0480266$). (b) Expansion of the final sequence of partial failures. (c) Further expansion of two partial failures.

Each element i is assigned a random time to failure t_{i0} under stress σ_0 based on Eq. (12). The actual time to failure of element i, namely T_{if}, is reduced below t_{i0} if stress is transferred to the element. The time T_{if} is obtained by requiring that Eq. (14) is satisfied.

Consider the example illustrated in Figure 4(a). The four elements $i = 1, 2, 3, 4$ carrying stress σ_0 are assigned failure times t_{10}, t_{20}, t_{30}, and t_{40} using the probability distribution Eq. (14). Element "2" has the shortest failure time so that

$$T_{2f} = t_{20}. \tag{15}$$

Upon failure the stress σ_0 carried by this element is transferred to element "1" as illustrated in Figure 4(a) (ii). Element "1" is the next element to fail and its failure time T_{1f} is given by

$$T_{1f} = t_{20} \left(1 - 2^{-\rho}\right) + 2^{-\rho} t_{10}. \tag{16}$$

Upon the failure of "1", the stress $2\sigma_0$ is transferred to elements "3" and "4" as illustrated in Figure 4(a) (iii). Element "4" is the next element to fail and its failure time T_{4f} is given by

$$T_{4f} = T_{1f} \left(1 - 2^{-\rho}\right) + 2^{-\rho} t_{40} = t_{20} \left(1 - 2^{-\rho}\right)^2 + t_{10} \left(1 - 2^{-\rho}\right) 2^{-\rho} + 2^{-\rho} t_{40}. \tag{17}$$

Upon the failure of "4," the stress σ_0 is transferred to element "3" as illustrated in Figure 4(a) (iv). The time to failure of element "3" is given by

$$T_{3f} = t_{20} \left(1 - 2^{-\rho}\right)^2 + (t_{10} + t_{40}) \left(1 - 2^{-\rho}\right) 2^{-\rho} + t_{30} 2^{-2\rho}. \tag{18}$$

Alternative failure sequences are also possible, one example is illustrated in Figure 4(b). Again "2" is the first element to fail, however in this case the second element to fail is "4," then "3" fails and finally "1" fails. Newman et al.[18] provide computationally optimal methods for performing the lifetime calculations for large arrays of elements and provide numerical evidence for the existence of a critical point.

RESULTS

We have carried out a sequence of numerical experiments using 12th-order ($N = 4,096$) and 16th-order ($N = 65,536$) realizations of our model with $\rho = 4$. An example of a 16th-order realization is given in Figure 6. The total failure sequence is given in Figure 6(a)). The nondimensional time is taken to be $\tau = \nu_0 t$ and failure in the case occurs at $\tau = 0.0480266$. It is interesting that failure occurs at a nondimensional time which is more that an order of magnitude *shorter* than the mean time to failure of an individual element $\tau_{1/2} = 0.61315$. The lifetime of our composite material is much shorter than the mean lifetime of individual elements.

This observation may help to explain why actual faults are much weaker than predictions based on laboratory friction experiments. Failure stresses on faults are about one order of magnitude lower than values obtained by extrapolations of the laboratory studies. This is known as the "heat flow paradox" since the expected frictional heating on faults is not observed.[14]

The failure sequence between $\tau = 0.0445$ and failure is expanded in Figure 6(b). There is a well-defined sequence of partial failures prior to the total failure at $\tau_f = 0.04802660$. Well-defined partial failures occur at $\tau_1 = 0.04796505$, $\tau_2 = 0.04779920$, $\tau_3 = 0.04748748$, $\tau_4 = 0.04716183$, and $\tau_5 = 0.04612445$. The failure sequence between $\tau = 0.04745$ and $\tau = 0.04785$ is further expanded in Figure 6(c) to show the structure of the partial failures at $\tau = 0.04779920$ and $\tau = 0.04748748$. In each case there are a nested sequence of higher-order partial failures. Further expansion would show higher orders of nesting. The structure is basically self-similar or fractal. There is a scale-invariant sequence of precursory failures at all levels. Because of the stochastic nature of the model the embedding is not always clear, a particular partial failure may be part of a sequence or may be precursory to another failure in the sequence. But this is also the problem with distributed seismicity.

It is also of interest to determine whether the sequence of partial failures can be inserted into the predictive log-periodic relation Eq. (5) in order to predict the time of the total failure. Taking the sequence of partial failures τ_5, τ_4, and τ_3 we obtain the prediction $\tau_f = 0.04763648$ from Eq. (5), taking the sequence τ_4, τ_3, and τ_2 we obtain $\tau_f = 0.05477475$ from Eq. (5), and taking the sequence τ_3, τ_2, and τ_1 we obtain $\tau_f = 0.04815362$ from Eq. (5). These results are summarized in Table 1. There is clearly considerable scatter in the predictions. Other realizations give similar results. Although the embedded sequences of precursory failures are a ubiquitous feature of all realizations, there is considerable stochastic variability of the timing. This is also a characteristic feature of distributed seismicity.

In Figure 7 the logarithm of the number of unfailed elements is given as a function of the logarithm of the time to failure for a realization with 4096 points and $\rho = 4$. The power-law fit shown by the dashed line has a slope of 0.24, this compares with the power $\rho = 0.25$ predicted by the global load sharing relation (11). Although there is considerable scatter, the power-law relation does appear to be a reasonable predictor for our model just as Bufe and Varnes[5] found for regional seismicity. It is important to note, however, that the quality of the fit deteriorates as complete failure is approached. The global analysis employed in the derivation of Eq. (11) deteriorates owing to the increasing importance of localization in the evolution of the cascade of failures.

The sequence of failures as a function of position on the linear array of elements is shown in Figure 8(a) for the above realization. An expanded version for the first 512 points is shown in Figure 8b. The precursory cascades of failure are clearly illustrated. This figure illustrates the growing importance of localization in failure events as criticality is approached. We will explore and analyze the behavior of the cascade when failure is imminent in a future paper.

TABLE 1 Table of Failure Times and "Predictions." Realization of failure using 65,536 point simulation. Three successive "events" employed to "predict" total failure. Three sets of successive events considered to provide estimate of error. True failure time also given.

$$\begin{bmatrix} 0.046125 \\ 0.047162 \\ 0.047487 \end{bmatrix} \Rightarrow \begin{array}{c} 0.047636 \\ \text{1st Est.} \end{array}$$

$$\begin{bmatrix} 0.047162 \\ 0.047487 \\ 0.047799 \end{bmatrix} \Rightarrow \begin{array}{c} 0.054775 \\ \text{2nd Est.} \end{array} \Rightarrow \begin{array}{c} 0.048627 \\ \text{True} \\ \text{Failure} \\ \text{Time} \end{array}$$

$$\begin{bmatrix} 0.047487 \\ 0.047799 \\ 0.047965 \end{bmatrix} \Rightarrow \begin{array}{c} 0.048154 \\ \text{3rd Est.} \end{array}$$

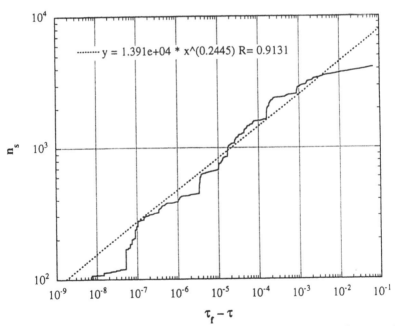

FIGURE 7 Failure sequence for a 12th-order ($N = 4,096$) realization of our model. The dashed line is a power-law fit to the failure sequence based on Eq. (5).

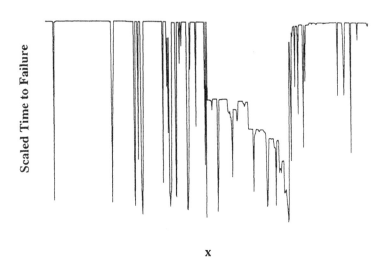

FIGURE 8   Sequence of failures as a function of position. First 512 elements.

CONCLUSIONS

We have explored by computational means the behavior of a hierarchical model for failure in time of a material. This hierarchical description accommodates the redistribution of stress from failed portions of a material to nearest regions of comparable size and provides a probabilistic realization of the accelerating tendency of material whose load has increased substantially to fail. This model also manifests some of the self-similar or "fractal" scaling features that are now recognized as being associated with seismicity and faulting. This class of models has been successful in earlier studies in describing the diminished threshold in the stress for failure ("fault weakening") as well as in the time, showing tantalizing hints for the existence of a critical point in the time domain. Building on that earlier work, we considered this model as a representation for seismic activation and find failure properties similar to those observed prior to large earthquakes. We plan to explore in future work the significance of the events occurring just before failure and their potential value in predicting great earthquakes.

ACKNOWLEDGMENTS

AMG was supported in part by the U.S. Army Research Office through the Army Center of Excellence for Symbolic Methods in Algorithmic Mathematics (AC-SyAM), Mathematical Sciences Institute of Cornell University, Contract DAAL03-91-C-0027.

REFERENCES

1. Aki, K. "A Probabilistic Synthesis of Precursory Phenomena." In *Earthquake Prediction*, edited by D. W. Simpson and P. G. Richards, 566–574. Washington, DC: American Geophysical Union, 1981.
2. Bak, P., C. Tang, and K. Wiesenfeld. "Self-Organized Criticality. *Phys. Rev.* **A38** (1988): 364–374.
3. Brown, S. R., C. H. Scholz, and J. B. Rundle. "A Simplified Spring-Block Model of Earthquakes." *Geophys. Res. Let.* **18** (1991): 215–218.
4. Bufe, C. G., S. P. Nishenko, and D. J. Varnes. "Seismicity Trends and Potential for Large Earthquakes in the Alaska-Aleutian Region." *PAGEOPH* **142** (1994): 83–99.
5. Bufe, C. G., and D. J. Varnes. "Predictive Modeling of the Seismic Cycle of the Greater San Francisco Bay Region." *J. Geophys. Res.* **98** (1993): 9871–9883.
6. Carlson, J. M., and J. S. Langer. "Mechanical Model of an Earthquake Fault." *Phys. Rev.* **A40** 6470–6484.
7. Carlson, J. M., J. S. Langer, B. E. Shaw, and C. Tang. "Intrinsic Properties of a Burridge-Knopoff Model of an Earthquake Fault." *Phys. Rev.* **A44** (1991): 884–897.
8. Coleman, B. D. "A Stochastic Process Model for Mechanical Breakdown." *Trans. Soc. Rheol.* **1** (1957): 153–168.
9. Coleman, B. D. "Statistics and Time-Dependence of Mechanical Breakdown in Fibers." *J. Appl. Phys.* **29** (1958): 968–983.
10. Hill, D. P., P. A. Reasenberg, A. Michael, W. J. Arabaz, G. Berooza, D. Brumbaugh, J. N. Brune, R. Castro, S. Davis, D. dePolo, W. L. Ellsworth, J. Gomberg, S. Harmsen, L. Howe, S. M. Jackson, M. J. S. Johnston, L. Jones, R. Keller, S. Malone, L. Munguia, S. Nava, J. C. Pechmann, A. Sanford, R. W. Simpson, R. B. Smith, M. Stark, M. Stizking, A. Videl, S. Walter, V. Wong, and J. Zollweg. "Seismicity remotely triggered by the magnitude 7.3 Landers, California, earthquake." *Science* **260** (1993): 1617–1623.
11. Huang, J., G. Narkounskaia, and D. L. Turcotte "A Cellular-Automata, Slider Block Model for Earthquakes II. Demonstration of Self-Organized Criticality for a 2-D System." *Geophys. J. Int.* **111** (1992): 259–269.

12. Ito, K., and M. Matsuzaki. "Earthquakes as Self-Organized Critical Phenomena." *J. Geophys. Res.* **95** (1990): 6853–6860.
13. Keilis-Borok, V. I. "The Lithosphere of the Earth as a Nonlinear System with Implications for Earthquake Prediction." *Rev. Geophys.* **28** (1990): 19–34.
14. Lachenbruch, A. H., and J. H. Sass. "Heat Flow from Cajon Pass, Fault Strength, and Tectonic Implications." *J. Geophys. Res.* **97** (1990): 4995–5015.
15. Nakanishi, H. "Cellular-Automaton Model of Earthquakes with Deterministic Dynamics." *Phys. Rev.* **A41** (1990): 7086–7089.
16. Nakanishi, H. "Statistical Properties of the Cellular-Automaton Model for Earthquakes." *Phys. Rev.* **A43** (1991): 6613–6621.
17. Newman, W. I., and A. M. Gabrielov "Failure of Hierarchical Distributions of Fibre Bundles I." *Intl. J. Fracture* **50** (1991): 1–14.
18. Newman, W. I., A. M. Gabrielov, T. A. Durand, S. L. Phoenix, and D. L. Turcotte. "An Exact Renormalization Model for Earthquakes and Material Failure. Statics and Dynamics." *Physica* **D77** (1994): 200–216.
19. Phoenix, S. L., and L. J. Tierney. "A Statistical Model for the Time-Dependent Failure of Unidirectional Composite Materials Under Local Elastic Load-Sharing Among Fibers." *Eng. Fract. Mech.* **18** (1983): 193–215.
20. Scholz, C. H. "A Physical Interpretation of the Haicheng Earthquake Prediction." *Nature* **267** (1977): 121–124.
21. Scholz, C. H. "Earthquakes and Faulting: Self-Organized Critical Phenomena with a Characteristic Dimension." In *Spontaneous Formation of Space-Time Structures and Criticality*, edited by T. Riste, and D. Sherrington, 41–56. Netherlands: Kluwer, 1991.
22. Smalley, R. F., D. L. Turcotte, and S. A. Solla. "A Renormalization Group Approach to the Stick-Slip Behavior of Faults." *J. Geophys. Res.* **90** (1985): 1894–1900.
23. Sornette, D., and C. G. Sammis "Universal Log-Periodic Correction to Renormalization Group Scaling for Regional Seismicity: Implications for Earthquake Prediction." Submitted for publication, 1995.
24. Sykes, L. R., and S. Jaume "Seismic Activity on Neighbouring Faults as a Long-Term Precursor to Large Earthquakes in the San Francisco Bay Area." *Nature* **348** (1990): 595–599.
25. Tse, S. T., and J. R. Rice "Crustal Earthquake Instability in Relation to the Depth Variation of Frictional Slip Properties." *J. Geophys. Res.* **91** (1986): 9452–9472.
26. Turcotte, D. L. *Fractals and Chaos in Geology and Geophysics.* Cambridge University Press, Cambridge, 1992.
27. Varnes, D. J. "Predicting Earthquakes by Analyzing Accelerating Precursory Seismic Activity." *PAGEOPH* **130** (1989): 661–686.
28. Zoback, M. D., and J. H. Healy. "In situ Stress Measurements to 3.5 km Depth in the Cajon Pass Scientific Research Borehole: Implications for the Mechanics of Crustal Faulting." *J. Geophys. Res.* **97** (1992): 5039–5057.

John B. Rundle,* William Klein,** S. Gross,* and Donald L. Turcotte†
*Department of Geological Sciences and CIRES, University of Colorado, Boulder, CO 80309
**Polymer Center and Department of Physics, Boston University, Boston, MA 02215
†Department of Geological Sciences, Cornell University, Ithaca, NY 14853

Observation of Boltzmann Fluctuations in Stochastic, Massless Slider-Block Simulations

Nonequilibrium lattice threshold models, of which slider-block models are an example, are now being used to model a variety of dynamical systems, including earthquake faults, driven neural networks, and sliding charge density waves. For a certain broad class of models, it can be shown that the internal energy field displays a Boltzmann (Gibbs) distribution. Numerical simulations confirm these predictions. Our results indicate that this class of models can be effectively treated as an equilibrium system, exhibiting detailed balance and "time reversal symmetry." Although slider-block models are highly idealized models for crustal seismicity, the universal behavior we have observed may well be applicable to the earth's crust and the distribution of earthquakes. Favorable evidence for this point of view comes from the near-universal applicability of power-law frequency-magnitude statistics to earthquakes, and the strong evidence for long-distance correlations between earthquakes.

INTRODUCTION

Earthquakes are one of the most feared of natural hazards. They strike without warning and often where least expected. Although the earth's crust is extremely complicated, the statistics of earthquake occurrence exhibit great simplicity. They universally satisfy the Gutenberg-Richter magnitude-frequency relation, which is a simple power law. There is also increasing evidence that earthquake occurrence is correlated over large distances. Consideration of the strain accumulation and release cycle on a single fault cannot simulate crustal deformation and earthquakes. The earth's crust must be treated as an interacting, self-organizing system capable of demonstrating critical phenomena.

Slider-block models[6,14,16,19] are simple examples of driven dynamical threshold systems. In addition to simulating aspects of earthquakes and frictional sliding, these models may also represent the dynamics of neurological networks[12] and sliding charge density waves.[4] In view of the importance of these nonequilibrium systems, a basic question can be asked as to whether one or more of these models exhibit any similarity to equilibrium systems. If such similarities exist, it is likely that some of the extensive body of results available for equilbrium systems might be useful in understanding the nonequilibrium threshold systems.

Although slider-block models can be considered grossly simplified models for the behavior of the earth's crust, they do exhibit some of the important characteristics of seismicity. They generally demonstrate chaotic behavior and, under a wide range of conditions, the frequency-magnitude statistics of slip events are power law. Many variations in slider-block models have been proposed during the past five years. The problems are greatly simplified if inertial effects are neglected, and if only nearest-neighbor interactions are considered. Slider-block models are in fact a subset of a broad class of lattice threshold models.

To summarize our main result: we have found broad classes of lattice models that possess an energy distribution approaching a Boltzmann-Gibbs (BG) distribution. These models display detailed balance, implying "time reversal symmetry." We test our prediction using nearest-neighbor slider-block simulations, and find that the agreement with the BG distribution improves as the model approaches mean field.

Consider a typical equilibrium system for which the total energy $E_T \approx$ constant, except for small fluctuations,[13] the mean-square probability of which decrease in magnitude as $1/\sqrt{N}$ where N is the number of particles (or degrees of freedom or modes). In such a model, the internal energy E_i of each independent field variable (molecules, spins, etc.) executes small fluctuations about the time-averaged mean energy. Assuming that the system obeys the postulate of equal *a priori* probability, the method of most probable distributions[13] can then be used to show that the expected distribution of block energies is Boltzmann. Dividing the possible energy states into $\{q = 1, \ldots, Q\}$ energy bands $E(q)$ occupied with probability $p(q)$, detailed balance requires that:

$$\Sigma_q p(E_q) = 1,$$
$$\Sigma_q p(E_q)E_q = E_T \approx \text{constant}. \tag{1}$$

Such systems will be symmetric under time reversal symmetry.[13] From Eq. (1) it is trivial[13] to show that the probability of a state $p(E_q) \propto \exp[-E_q/\varepsilon]$, the Boltzmann-Gibbs distribution, where ε is the lattice-average energy per degree of freedom ($\varepsilon =$ mean energy = "temperature"). It is well known that most nonequilibrium systems do not display detailed balance, are not symmetric under time reversal, and do not possess Boltzmann distributions. The lack of time reversal symmetry, and violation of the assumption of equal *a priori* probability, is reflected in macroscopic processes of dissipation.

MODEL

Slider-block models have been proposed as analogs for the stick-slip behavior of faults; however, these models also exhibit a wide range of interesting phenomena. Consider a two-dimensional array of N blocks with each block connected to its nearest neighbor by coupling springs with spring constant K_C, and to a loader plate by a loader spring with spring constant K_L. The loader plate translates at velocity V, increasing the force on each sticking block until it slips when the force (stress) exceeds the static friction. With a static friction greater than the dynamic friction, or with velocity weakening friction, stick-slip behavior is observed.

In its most general form, the model is completely deterministic and the equations of motion for the slipping blocks must be solved simultaneously. The model is greatly simplified if only one block is allowed to slide at a time, this is then a cellular automaton (CA) slider-block model.[14,19] The model can be further simplified by neglecting the mass of the blocks. In this case, the force threshold for slip σ_i^F and the residual force after slip are specified. Blocks move when the force on the ith block σ_i equals or exceeds the threshold. The CA models are distinguished from the original massive Burridge-Knopoff model[6] in that the position of each block through time is obtained from an update rule, instead of by solving a set of N coupled differential equations. Farther-neighbor models can also be implemented to represent elastic continua[16] using long-range coupling springs whose spring constants decay with distance r as $1/r^3$. An advantage of massless CA models is that the dynamics of large N models can be examined on even modest workstations, which can be important when correlation lengths are large and finite size effects are important. Moreover, CA models can be used to simulate driven integrate-and-fire neural networks, by identifying the lattice slip variable with neural cell voltage, and the failure threshold with the cell firing voltage.[12]

All massless slider-block models in the literature oscillate around a fixed value of energy. A subset of these models are deterministic, and are known to have periodic limit cycles.[9,10] Another class of models,[15] also deterministic, do not have a limit cycle attractor, but fluctuate about a constant energy. These models are candidates for BG models, although this result has as yet not been demonstrated. The major focus here is a third and separate class[5] of slider-block models that have a stochastic dynamics, and which can be shown to possess a BG distribution. These models may have utility as either viable candidates for understanding the physics of earthquakes, or at the least may find use in illuminating features of other classes of deterministic slider-block models.

A Hamiltonian[16,19] for slider-block models with arbitrary range interactions having coupling springs of fixed strength K_C can be written in the form:

$$H_{sb} = \left(\frac{1}{2}\right) \Sigma_i \left\{ K_L \phi_i^2 + \left(\frac{1}{2}\right) K_C \Sigma_{j=\text{int}} [\phi_j - \phi_i]^2 \right\}$$
$$= \exp\left(\frac{1}{2}\right) \Sigma_i \left\{ K_T \phi_i^2 - 2K_C \Sigma_{j=\text{int}} \phi_i \phi_j \right\}$$

$$(2)$$

where K_T is the total spring constant ($K_T = K_L + 2dK_C, d = $ dimension of space). The sum over i is over all sites in the lattice, and the sum over j is over all interacting blocks within the range of interaction R but excluding site i. The order parameter field is the slip deficit $\phi_i(t) = s_i(t) - Vt, s_i(t)$ is the slip of block i at time t. The value of H_{sb} for a particular configuration $\{\phi_i\}$ is $E_{sb} = H_{sb}(\phi)$. The force (stress) σ_i on block i is:

$$\Sigma_{sb,i} = -\frac{\partial H_{sb}}{\partial \phi_i} = -\{K_L \phi_i + K_C \Sigma_{j=\text{int}} [\phi_j - \phi_i]\}.$$

$$(3)$$

Expression (2) can be formally obtained[16,19] from the expression for the elastic strain energy of a fault embedded within an elastic continuum by the use of a gradient expansion. For the nearest-neighbor models, the spring constants K_L and K_C are then related respectively to the zeroth and second moments of the stress Green's function. Note that the Hamiltonian (2) is identical to the Gaussian model[4] for spin systems. If we were to constrain the slider-block model to have:

$$\Sigma_i \phi_i^2 = \text{constant},$$

$$(4)$$

then Eqs. (3) and (4) would be identical to the spherical model.[2] One expects that a condition similar to Eq. (4) may hold approximately in the steady state for some deterministic slider-block models with a threshold, inasmuch as the values of ϕ_i are effectively constrained to lie within bounds arising from the parameter values.

These slider-block models may share therefore some similarities with the spherical model.

We now introduce a small modification[17] of Eq. (2) for dynamical calculations in which a loader plate time scale $\tau \to 0$ exists. The physical reason for this modification arises from the condition that energy must be stored by the system before slip of a block can occur. This modified Hamiltonian H is:

$$H = \left(\frac{1}{2}\right) \Sigma_i \left\{ K_L(\phi_i - V\tau)^2 + \left(\frac{1}{2}\right) K_C \Sigma_{j=\text{int}} [\phi_i - \phi_i]^2 \right\} ; \tag{5}$$

i.e., Eq. (5) is just Eq. (2) with ϕ_i replaced by $\phi_i - V\tau$. Then the stress is:

$$\sigma_i = -\frac{\partial H}{\partial \phi_i} = -\{K_L\phi_i + K_C\Sigma_{j=\text{int}}[\phi_i - \phi_i]\} + K_L V\tau. \tag{6}$$

The first term in brackets is the same as Eq. (3), the last term is the stress that is produced on one loader plate update and can be thought of as a "prestress." For convenience, we define the energy E_i of the ith block by the expression $H = \Sigma_i E_i$. For the moment, we further specialize to models with $V \to 0$, in which only one avalanche occurs following each loader plate update, although the difference between Eqs. (2) and (5) becomes important in deriving the associated Langevin equation.[17]

For CA models, a rule to generate the dynamics must be specified. The simplest example is the modified Mohr-Coulomb friction law, in which each block has a prescribed static failure threshold σ_i^F, and a residual stress at which the block sticks, σ_i^R. The dynamics are generated by a jump rule giving the position of the block as a function of the state of stress on the block. Jump rules such as these can be either deterministic or stochastic. The basic jump rule is:

$$s_i(t + 1) = s_i(t) + J(\sigma_i)\Theta(\sigma_i - \sigma_i^F) \tag{7}$$

where $\Theta(x)$ is a Heaviside step, and the failure threshold σ_i^F is spatially dependent. Examples[16] of deterministic jump functions $J(\sigma_i) = \Delta s_i$ include:

$$J_1 = \frac{\sigma_i - \sigma_i^R}{K_T},$$
$$J_2 = \frac{\sigma_i^F - \sigma_i^R}{K_T}, \tag{8}$$

where σ_i^R is a residual stress. This expression is valid also for a model with longer range springs (interactions), in which each block interacts with N_R other blocks via springs with spring constants K_C. Each block jumps from its current stress at failure to the position having the specified residual stress σ_i^R, thus Eqs. (7)–(8) is an example of a deterministic rule.

Stochastc models are also possible, and one of these[5] is used in the simulations here. For this model, the jump is given by:

$$J_s = J_1(1 - W\rho) \tag{9}$$

where W is a (constant) width chosen from $0 \leq W \leq 1$, and ρ is a uniformly distributed random number on $\rho \in [0, 1]$.

To understand the results of the simulations carried out below, we observe that the slip deficit $\phi_i(t)$ of a block fluctuates around a time-averaged value η_i:

$$\eta_i = \frac{1}{T} \int_0^T \phi_i(t)dt = \{\phi_i(t)\}. \tag{10}$$

The fluctuating part $\Psi_i(t)$ is then defined by:

$$\phi_i(t) = \eta_i + \Psi_i(t). \tag{11}$$

For convenience in analyzing data obtained in simulations using the jump (9), we normalize the time-average fluctuation $\Psi_i(t)$ to unity by defining the variance ω^2 of the fluctuation:

$$\omega_i^2 = \frac{1}{T} \int_0^T [\Psi_i(t)]^2 dt = \{[\Psi_i(t)]^2\}. \tag{12}$$

We have:

$$\Psi_i'(t) = \frac{1}{\omega_i} \Psi_i(t). \tag{13}$$

For simulations in which $V \to 0$, the Hamiltonian in Eq. (5) can now be written as:

$$H = H_0 + H_1 + H' \tag{14}$$

where the various terms are defined as:

$$\begin{aligned}
H_0 &= (1/2)\Sigma_i\{K_L(\eta_i)^2 + (1/2)K_C\Sigma_j[\eta_j - \eta_i]^2\}, \\
H_1 &= \Sigma_i\{\omega_i K_L \eta_i \Psi_i' + (1/2)\omega_i K_C \Sigma_j[\gamma_{ij}\Psi_j' - \Psi_i'][\eta_j - \eta_i]\}, \\
H' &= (1/2)\Sigma_i \omega_i^2\{K_L(\Psi_i')^2 + (1/2)K_C\Sigma_j[\gamma_{ij}\Psi_j' - \Psi_i']^2\},
\end{aligned} \tag{15}$$

and where:

$$\gamma_{ij} = \frac{\omega_j}{\omega_i}. \tag{16}$$

Our goal is to accumulate time-averaged statistics of the energy distribution corresponding to the energy given by Eq. (14). Taking the time average of $\{H\}$ in Eq. (14), we observe that:

$$\begin{aligned}
\{H_o\} &= H_o, \\
\{H_1\} &= 0, \\
\{H'\} &\neq 0.
\end{aligned} \tag{17}$$

H_o is by definition a constant, H_1 executes small fluctuations about 0, and H' fluctuates about some nonzero value. Since we are interested in the nonzero time-averaged occupation numbers n_q for the various energy bands centered on E_q, we focus attention on the last of Eq. (17). We therefore define:

$$\{H'\} \equiv \Sigma_q n_q E_q . \tag{18}$$

By comparison of Eq. (18) with Eqs. (1)–(2), we surmise that the numbers n_q of blocks that occupy energy levels E_q should display Boltzmann-type statistics, assuming that all configurations have equal *a priori* probability. In fact, the probability density functions (PDF) of individual springs should be Gaussian in Ψ. The total energy of ν identical independent springs is then chi-square distributed with ν degrees of freedom,

$$(E_q/\varepsilon)^{(\nu/2-1)} e^{-E_q/\varepsilon} ,$$

where ε is the mean energy. This is the hypothesis we test with our simulations.

In analogy with well-known results arising from the Boltzmann transport theorem in kinetic theory,[13] we expect that the property of microscopic chaos will exist in the slider-block models when the interaction between neighboring blocks is large enough to self-organize the blocks against the competing stochastic noise from the block jumps. In turn, some level of stochastic noise must be present, particularly for small values of K_C, to prevent the blocks from phase-locking into a limit cycle.[9,10] Since $R = (K_C/K_L)^{1/2}$ is a measure of the range of interaction, models with large values of R display mean-field characteristics. Mean-field models with a given noise level are more likely to demonstrate Boltzmann statistics: these models are associated with decreasing amplitude fluctuations at all but the largest wavelengths; thus the assumption $\Sigma_q n_q E_q = $ constant is more likely to be valid.

SIMULATIONS

We have carried out a number of stochastic simulations on a square lattice with nearest-neighbor interactions using Eqs. (6)–(9). For an infinitely large lattice, $\eta_i = \eta = $ constant, and $\omega_i = \omega = $ constant, $\gamma_{ij} = 1$, but the presence of finite boundaries causes η_i and ω_i to vary across the lattice. For that reason, we accumulate time-averaged statistics by using the normalized energies:

$$H_i'' = (1/2)\{K_L(\Psi_i')^2 + (1/2)K_C \Sigma_j [\Psi_j' - \Psi_i']^2\} \tag{19}$$

to construct the time-averaged, cumulative distribution function for the block energies, and to plot the lattice-averaged energy against time. Operationally, we define

10,000 energy bands centered on each E_q, $[q = 1, \ldots, 10000]$, and, upon termination of all block motion, count the number of block energies falling into each narrow energy band following a loader plate update . Defining:

$$p(E_q) = \frac{n_q}{N} \tag{20}$$

as the probability density function, we see that the probability of a block being in the energy band centered on E_q, so that it can be counted, is $E_q p(E_q)$. If $p(E_q)$ are normalized Boltzmann functions:

$$p(E_q) = \frac{1}{\varepsilon} \exp[-E_q/\varepsilon], \tag{21}$$

the cumulative distribution function $P(E')$ is obtained from:

$$P(E') = \int_0^{E'} E' \exp[-E'] dE'$$
$$= 1 - (1 - E') \exp[-E'] \quad , \quad E' = \frac{E_q}{\varepsilon} . \tag{22}$$

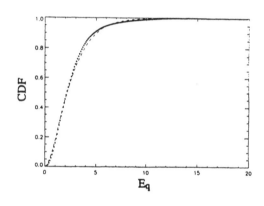

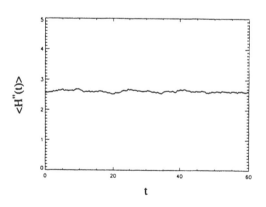

FIGURE 1   (Upper) Cumulative distribution of block energies equation from our simulation (dots) and prediction (dashed line) using Eqs. (21)–(22). (Lower) Lattice average of the block energies $\langle H''(t) \rangle$ as a function of time. Mean energy ε used in constructing the dashed line in the upper figure is obtained by averaging $\langle H''(t) \rangle$ over the time interval of the simulation, the average value obtained being 2ε. Parameters in this simulation are $K_C = 1, K_L = 1, W = .8$. Time-averaged energy $\epsilon = 3.605, \sigma^F = 35, \sigma^R = 0$.

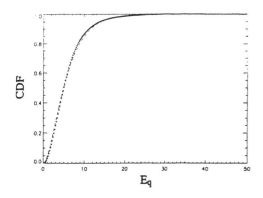

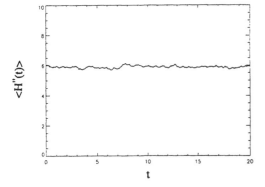

FIGURE 2 Same as Figure 1 with $K_C = 5, K_L = 1, W = .7$. Time-averaged energy $\varepsilon = 2.944, \sigma^F = 35, \sigma^R = 0$ here.

Defining the lattice average of the block energies H_i'' at fixed time t by $\langle H''(t) \rangle$, we obtain the time-averaged energy per block ε (= "temperature") as:

$$\varepsilon = (1/2)\{\langle H''(t) \rangle\}. \tag{23}$$

Equations (22)–(23) represent a firm prediction, with no free parameters, of the energy distribution obtained from simulation data. Three examples of our simulations are given in Figures 1–3. Simulations were carried out on a 100×100 square lattice of points, and in all figures, $K_L = 1$. Values of K_C range from $K_C = 1$ (Figure 1), $K_C = 5$ (Figure 2), to $K_C = 50$ (Figure 3). Values for W were $W = .8$ (Figure 1), $W = .7$ (Figure 2), and $W = .1$ (Figure 3).

For each case, the lattice-averaged energies $\langle H''(t) \rangle$ are found at the time of each avalanche cluster. These values, which fluctuate with time, are shown in the bottom panel of each figure. The lattice-averaged energies $\langle H''(t) \rangle$ shown in the figures are then averaged over time to obtain the "temperatures" ε for each simulation, to be used together with Eq. (22) in calculating the CDF, the dashed curve in the top panel of each figure. These dashed, theoretical CDF curves are then compared with the experimentally determined CDF measured from the simulations (dots). Agreement between theory and simulation data is good in Figures 2 and 3, somewhat less so in Figure 1. As expected, models that are closer to mean field (Figures 2 and 3) are better represented by Boltzmann statistics. In fact, it can be shown that all mean-field slider-block models have Boltzmann energy distributions.[13,18] Because the line of reasoning does not depend either on the massless nature of the slider blocks, we predict that similar results will be observed in massive slider-block simulations[6,8] as well. Since the noise amplitude required to generate the BG distribution decreases as mean field is approached, we also predict that the amplitude of the external noise should be vanishingly small in the mean-field limit.[18,7]

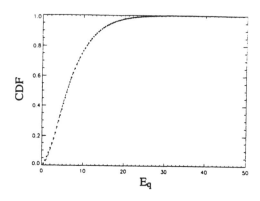

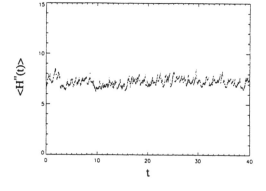

FIGURE 3 Same as Figure 1 with $K_C = 50, K_L = 1, W = .1$. Time-averaged energy $\varepsilon = 1.305, \sigma^F = 35, \sigma^R = 0$ here.

DISCUSSION

Clearly the model considered here is highly idealized relative to the earth's crust. However, we believe that the demonstration that a dissipative complex system exhibit Boltzmann fluctuations has far-ranging implications. We expect that the results demonstrated here will be applicable to a wide range of interacting, nonlinear lattice threshold models.

In terms of distributed seismicity, the results of this chapter are quite significant. Earthquakes can be interpreted as correlated statistical fluctuations in an interacting system, the earth's crust, providing a fundamental explanation for the wide applicability of the Gutenberg-Richter frequency-magnitude statistics. In addition, the concept of correlated fluctuations provides a basis for accepting the correlations of seismic activity observed over large distances.[11]

A question that must be considered is whether the occurrence of very large earthquakes are statistically correlated over very large distances, even the entire earth. The concepts introduced in this chapter suggest that this can be possible. If the entire crust is in an elevated energy state as a result of a random fluctuation (high stress), then the occurrence of large earthquakes could be expected throughout the system.

ACKNOWLEDGMENTS

Work carried out by JBR was supported under U.S. Department of Energy grant number DE-FG03-95ER14499 to the Cooperative Institute for Research in Environmental Sciences at the University of Colorado. The authors would also like to acknowledge the generous hospitality of the Santa Fe Institute where much of this work was carried out.

RREFERENCES

1. Ben-Zion, Y., and J. R. Rice. *J. Geophys. Res.* (1995): to appear.
2. Berlin, T. H., and M. Kac. *Phys. Rev.* **86** (1952): 821.
3. Binney, J. J., N. J. Dowrick, A. J. Fisher, and M. E. J. Newman. *The Theory of Critical Phenomena.* Oxford: Clarendon Press, 1992.
4. Brown, S., and G. Gruner. *Sci. Am.* **270** (1994): 50.
5. Brown, S. R., C. H. Scholz, and J. B. Rundle. *Geophys. Res. Lett.* **18** (1991): 215–218.

6. Burridge, R., and L. Knopoff. "Model and Theoretical Seismicity." *Bull. Seism. Soc. Am.* **57** (1967): 341.
7. Carlson, J. M., and J. S. Langer. *Phys. Rev. A* **40** (1989): 6470.
8. Ferguson, C., W. Klein, and J. B. Rundle. Unpublished manuscript, 1995 (to be published).
9. Gabrielov, A., W. Newman, and L. Knopoff. *Phys. Rev. E* **50** (1994): 188–197.
10. Herz, A. V. M., and J. J. Hopfield. Preprint, 1994.
11. Hill, D. P. *Science* **260** (1993): 1617–1623.
12. Hopfield, J. J. *Physics Today* **47** (1994): 40.
13. Huang, K. *Statistical Mechanics*. New York: John Wiley, 1987.
14. Rundle, J. B., and D. D. Jackson. "Numerical Simulation of Earthquake Sequences." *Bull. Seism. Soc. Am.* **67** (1977): 1363.
15. Rundle, J. B., and W. Klein. *J. Stat. Phys.* **72** (1993): 405.
16. Rundle, J. B., and D. L. Turcotte. "New Directions in Theoretical Studies of Tectonic Deformation." In *Contrib. Space Geodesy to Geodynamics: Crustal Dynamics*, edited by D. E. Smith and D. L. Turcotte. Geodynamics Ser. Vol. 23, Washington, DC: AGU, 1993.
17. Rundle, J. B., W. Klein, and D. L. Turcotte. *Phys. Rev. Lett.* (1995): submitted.
18. Rundle, J. B., W. Klein, S. Gross, and D. L. Turcotte. *Phys. Rev. Lett.* **75** (1995): 1658–1661.
19. Rundle, J. B., W. Klein. and S. Gross. This volume.
20. Van Kampen, N. G. *Stochastic Processes in Physics and Chemistry*. Amsterdam: North-Holland, 1981.

J. M. Carlson,[†] S. L. Pepke,[†] and V. Kossobokov[‡]
[†] Department of Physics, University of California, Santa Barbara, CA 93106
[‡] International Institute for Earthquake Prediction Theory and Mathematical Geophysics, Russian Academy of Sciences, Warshavskoye Shosse, 79 k. 2, Moscow 113556

Prediction Studies of Earthquake Fault Models and Applications to Seismic Catalogs

We present an overview of our ongoing studies of the predictability of large events in a collection of dynamical models of seismicity. We focus on methods that are based on the detection of specific patterns within the catalog of events. In all of the models we have considered there is at least some correlation between precursory smaller events and the large events we aim to forecast, although in some cases the correlation is extremely weak. In the context of the models, we assess the relative effectiveness of different precursors, and find that a new measure called active zone size (AZS), which counts the fraction of the total seismic area that has exhibited earthquakes recently, is more strongly correlated with large events than the more conventional measure activity (N), which is defined to be the total number of earthquakes. We also discuss applications of AZS to real catalogs. We compare measurements of AZS and N both individually and in the context of the intermediate-term earthquake prediction algorithm M8 applied to data from the western United States. While neither measure considered alone can be used to reliably forecast large earthquakes, when we replace functions in M8 which are based on N with similar functions based on AZS, we have found that at least on a limited data set the performance of M8 is somewhat improved.

Reduction & Predictability of Natural Disasters, Eds. Rundle, Turcotte, & Klein,
SFI Studies in the Sciences of Complexity, Vol. XXV, Addison-Wesley, 1996   **273**

1. INTRODUCTION

Compared to many other natural hazards (e.g., hurricanes, floods, tornados), predicting large earthquakes poses a particularly challenging problem. In many of the other cases, it is possible to detect the hazard as it develops. For example, hurricanes can be monitored as they develop in the ocean, and the prediction problem becomes one of estimating when and where the hurricane will impact the coast. In such cases, a good prediction allows emergency measures to be taken and populations to evacuate in regions which are likely to suffer the most damage. In contrast, to date no clear indication of imminent large-scale seismicity has been found. This may be due in part to observational difficulties. The time scales on which tectonic stresses build are of order hundreds of years or more and the spatial regions of development are large, prohibiting continuous detailed monitoring. In addition, most of the deformation is hidden underground, so that there is no direct method of measuring the current state of stress. Finally, once a large rupture begins, the duration of the earthquake is very short (minutes or less), and damage takes place essentially immediately, so that unlike the hurricane there is simply insufficient time to begin emergency measures such as evacuation.

For these reasons, earthquake predictions fundamentally involve much greater uncertainty. Typically, predictions are classified by the time scale on which they are thought to apply, and many methods (though certainly not all methods) are based on information which is contained in earthquake catalogs. The basic goal is to determine the probability of a large earthquake within some time window extending into the future, given a catalog listing the times, locations, and sizes of events in the past, so that prediction is essentially a statistical question. For example, the goal of long-term prediction is to estimate the probability a large earthquake will occur in a given region within roughly 30 years,[22] and these assessments are typically based on the magnitudes of previous large earthquakes in the region and the time intervals between them. The goal of intermediate-term prediction is to assess the hazard on a time scale of roughly a few years, and methods are based on features such as increases in the rate of smaller earthquakes in the region and other precursory phenomena.[12,13] Finally, the goal of short-term prediction is to assess the hazard on a time scale of days or months, and some of these methods are based on observations of immediate foreshocks.[11] While prediction methods spanning all of these time scales are currently being investigated, the results are difficult to evaluate. Only a limited amount of data is currently available, and the time scales on which data must be gathered in the earth are sufficiently long that detailed quantitative tests of these kinds of forecasting methods are not likely to occur in our lifetime.

Consequently, earthquake models are beginning to play an important role in the development and testing of prediction techniques. These models are complex spatially extended dynamical systems, which yield rich (typically chaotic) patterns of behavior consisting of sharply defined individual events. Synthetic catalogs generated by models can provide the equivalent of millions of years of data, and are perhaps the only hope for systematic assessment of prediction methods. This is not because models behave exactly like the earth. Rather models open up the possibility of optimizing on a clean system, leading to new algorithms and estimates of intrinsic limitations and uncertainties. The combined effort of studying both models and seismic data should ultimately lead to the development of more realistic models as well as better algorithms for prediction.

In this paper we discuss some of our recent results on prediction studies of models and studies of seismicity in the earth. We focus on results which pertain to intermediate-term prediction techniques, in particular those based upon the pattern recognition algorithm M8 which was introduced by Keilis-Borok and Kossobokov[13] as a possible means to objectively predict large earthquakes worldwide. In Section 2 we outline the basic prediction method. In Section 3 we discuss applications of this method to predicting large events in models. In Section 4 we present some recent applications of what we have learned from model studies to seismic data sets. Finally, in Section 5 we discuss some of the implications that the model studies and seismic data studies have for each other, which point toward directions for future research.

2. INTERMEDIATE-TERM PREDICTION TECHNIQUES

Our method of forecasting is based on the pattern recognition algorithm M8 introduced by Keilis-Borok and Kossobokov[13] which is a candidate method for intermediate-term prediction of large earthquakes. The method is formulated on the hypothesis that regional small-scale seismicity may be used to diagnose an upcoming large event. The first step is to coarse-grain space into a set of overlapping circles, referred to as regions of investigation. In applications to the earth, the sizes of the circles are set by the lower magnitude cutoff of the target events to be predicted; the diameter of each circle is taken to be an order of magnitude larger than the estimated (minimum) length of the rupture. Then within each spatial region a set of functions is evaluated based on the earthquakes which occur in the region within a specified time window. These functions include the activity N, which within each space-time window is defined to be: $N = \#$ earthquakes. The activity weights all earthquakes equally, regardless of their size, and is therefore dominated by the smallest events which are most numerous. Note that for seismic catalogs, N must be defined with respect to some lower magnitude cutoff to insure completeness of the record. In fact, in M8 a total of four of the measures are direct variations of N, obtained by incorporating different lower magnitude cutoffs for the count and comparing the current value of N to the long time average. In this way the basic measure N can be used to monitor the rates of events in different size ranges, as well as fluctuations in these values. Prior to a large event the most effective measures will often take elevated (or depressed) values relative to the average value.

In M8 instead of forecasting precisely when and where an event is likely to take place, one issues a "time of increased probability" (TIP) within a particular circle of investigation when it is thought to be in a state for which a large event is likely. The goal is to maximize the rate of successful TIPs, while simultaneously minimizing the associated alarm time. In the earth no single measure has yet been identified that reliably predicts all of the large events. In M8 seven earthquakes, precursors are combined in a voting algorithm: if a fixed number of the precursors exceed individual specified thresholds, then the TIP is turned on and maintained for a time period of five years after which the prediction is reevaluated. Note that there are many specific parameters and functions that are intrinsic to the definition of M8. Examples include the seven chosen precursors, the centers and sizes of the circles of investigation, and the five-year alarm duration. However, the settings described here do not represent the result of a detailed optimization, but instead are selected because it is thought that they represent a solid first approximation to some future optimal version of the method.

The first test of M8 was based on existing data, and alarms were maintained for 20% of the test period and led to successful predictions for 80% of the large events. More recently, an ongoing test for predicting future events in the Circum Pacific has been established,[10] and efforts have been made to reduce the sensitivity of the results to parameters that are inherent in the algorithm,[15] and uncertainties and errors in the catalogs.[9] While it is still not clear how well this method can be made to perform on seismic data, it stands out as being the most objective algorithm currently under consideration for assessing large earthquake probabilities worldwide.

While we will return to the original form of M8 when we discuss applications to seismic data, in the context of the models, we consider a simplified version of the algorithm. In the simplified version, precursors are considered individually, so that we can directly assess the relative performance of different functions on the catalog of events. In addition, we do not specify an *a priori* fixed time window over which the prediction is maintained. Instead, precursors are reevaluated after each event (in most cases the model dynamics are such that alarms are maintained until a large event once they have been activated). We turn on a TIP when the precursor exceeds a threshold, and turn the TIP off when the precursor falls below the threshold. Thus by varying the threshold we vary the total alarm time, and from this we construct a *success curve*[16] which illustrates the fraction of events predicted as a function of the fraction of the the total space-time volume in which TIPs are turned on (some examples are illustrated in Figure 3). Note that the 45° line illustrates the results obtained when alarms are issued on a purely random basis, or the results that would be obtained for any prediction technique (with a variable alarm time) if the system that is studied behaves in a purely random fashion. Thus deviations from this line, whether above or below, give a measure of predictability of the system and the performance of different algorithms. When a precursor tends to exhibit depressed rather than elevated values prior to large events, we will illustrate success curves for the complement of the original function (obtained by taking the negative of the original function and threshold value, and issuing alarms when the negative function exceeds the negative threshold).

In the next section we use success curves to compare the effectiveness of different precursors and the predictability of different models. Because we can generate essentially arbitrarily long catalogs for the models, we can insure that the results we obtain have converged to their asymptotic limits. For seismic studies there are simply too few large events to even begin to plot the results on a success curve. Instead the threshold values are *a priori* fixed in M8 to be the top 10% (and the top 25% for one of the measures) of values that have been observed to date, and each large event results in a score in favor of or against M8 as a tool for intermediate-term prediction.

3. PREDICTION STUDIES OF EARTHQUAKE FAULT MODELS

We apply the intermediate-term prediction technique to four models which have been suggested as possible dynamical analogs of seismic phenomena. We refer to them as theChen-Bak-Obukhov model the Bak, Tang, and Wiesenfeld (BTW) model,[1] the Olami, Feder, and Christensen (OFC) model,[17] the Chen, Bak, and Obukhov (CBO) model,[7] and the Uniform Burridge and Knopoff (UBK) model.[2,3,4] While the first three of these (BTW, OFC, and CBO) differ from one another in certain important ways, at least for the system sizes considered here, they all generate pure power-law event-size distributions:

$$P(s) = s^{-(b+1)} \tag{1}$$

as illustrated in Figure 1. This distribution is analogous to the Gutenberg–Richter law[8] describing seismicity catalogs taken from the entire earth or large regional fault systems. These three systems are all examples of self-organized criticality (SOC), a concept introduced by Bak, Tang, and Wiesenfeld,[1] which is a theory that attempts to explain the broad distribution of events as well as the geometric complexity of fault networks within a dynamical framework. Here the emphasis on "criticality" because the power-law event-size distribution extends from the smallest event size, up to essentially the system size. While the origin of complex behavior in the fourth model is also a dynamical instability, the UBK model is not critical. The event size distribution consists of a power law describing the small to moderate events and excess large events, which cut off at some characteristic size, independent of the system size[6] for systems that are large enough (Figure 1). This statistical distribution is analogous to what is thought to occur for individual faults or narrow fault zones, where the largest events dominate the total slip, and occur at a rate that exceeds the extrapolated rate of small to moderate events.[21]

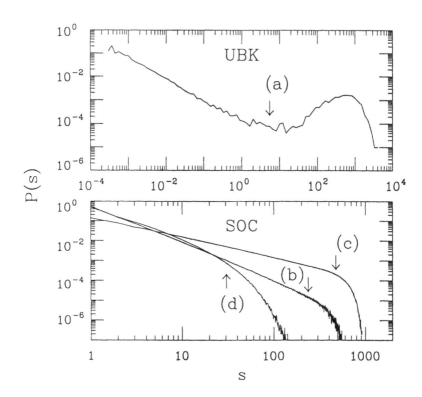

FIGURE 1 Event size distributions $P(s)$ vs. s for the (a) UBK, (b) OFC, (c) BTW, and (d) CBO models. In each case s is a measure of the size of the event--the integrated slip (seismic moment) for the UBK model, and the number of sites that topple for the others. In each case we attempt to predict events with $s \geq \tilde{s}$, where $\tilde{s}$ is some characteristic size (in each case $\tilde{s}$ is marked with an arrow). We take: $L = 8192$, $\sigma = .01$, $\alpha = 3$, and $\xi/a = 10$ for the UBK model,[18] system sizes 32×32 for the other models, and $\alpha = .2$ for the OFC model.[17]

We begin with the UBK model[3,4] for which the fault displacement $U(x,t)$ as a function of position x and time t satisfies a nonlinear wave equation

$$\frac{\partial^2 U}{\partial t^2} = \frac{\partial^2 U}{\partial x^2} - U - \phi(\dot{U}) + \nu t \tag{2}$$

in which ν represents the slow steady driving rate which we take to be infinitesimally small. The key instability leading to chaotic behavior is a velocity-weakening, stick-slip friction law

$$\phi(\dot{U}) = \begin{cases} (-\infty, 1], & \dot{U} = 0 \\ \dfrac{(1-\sigma)}{1+(2\alpha\dot{U}/(1-\sigma))}, & \dot{U} > 0, \end{cases} \tag{3}$$

in which σ and α are parameters. In the finite difference approximation, the model can be thought of as a one-dimensional chain of blocks, which is pulled slowly across a rough surface. Eventually some block reaches the sticking friction threshold and begins to slide, possibly destabilizing a string of neighboring blocks as a consequence of its motion. The very small driving rate leads to a separation of time scales between the duration of individual events and the time intervals separating consecutive events. Starting with some small inhomogeneity, after an initial transient the system reaches a statistically steady state characterized by the event-size distribution illustrated in Figure 1. Note that there is a sharp distinction between small ($s < \tilde{s}$) and large ($s \geq \tilde{s}$) events, and it is our goal is to predict the large events.

A segment of the catalog of events as a function of space and time is illustrated in Figure 2. For each event, a line is drawn through all the blocks that slip and a cross marks the epicenter of each large event. Precursory small-scale seismicity appears to be correlated with future large events; however, it begins on average after only half of the mean recurrence interval between large events has passed.

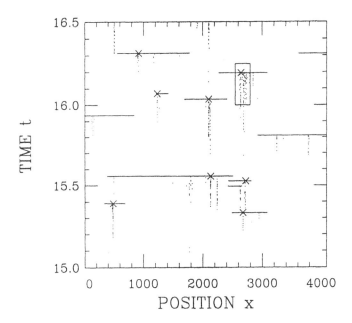

FIGURE 2   A small sample catalog as a function of space x (block number) and time t in the UBK model. A line segment marks blocks that slip in each event, and a cross marks the epicenter of each large event, which is clearly correlated with the small-scale seismicity. The box corresponds to a space-time window within which A and AZS are evaluated.

Thus from Figure 2 alone it is not clear how accurate predictions based on this pattern[20] will be.

In Pepke et al.,[18] a detailed study of predictability of the UBK model was performed. The system was coarse-grained into a set of overlapping regions (in this case, optimal performance was obtained when the regions were taken to be somewhat smaller than the mean size of large events), and a variety of different precursors were compared. This study revealed that among the set of measures considered, the two most effective precursors are activity N and a new spatial measure which we call active zone size AZS, which for each space-time window is defined to be: $AZS = \#$ blocks that have slipped (independent of the number of times). With N we are able to predict 90% of the large events, with alarms that occupy 15% of the space-time volume (Figure 3). With AZS we successfully predict 90% of the large events when alarms occupy only 8% of the space-time volume. In the UBK model the effectiveness of AZS can be traced to the fact that very little stress is relieved when a block slips in a small event. Instead small events serve as markers that the region is locally close to threshold. While the two precursors are clearly not independent, in comparison to N, AZS is a more direct measure of the size of the region that is near the threshold for slipping, and thus ultimately leads to the more direct assessment of the probability of a large event.

Next we compare these two precursors in the context of the three SOC models.[19] For each of these the system can be thought of as a two-dimensional[5] lattice of "blocks" with open boundary conditions. Unlike the UBK model these models ignore the details of inertial dynamics and friction laws, and instead evolve according to specified "breaking rules," so that when the stress $h(i, j)$ of a local block exceeds a threshold, it relaxes according to some avalanche dynamics. In each case there is a particular rule that specifies the stress drop of the toppling site, the increases in stress of other sites, and the net stress drop of the system. Most importantly, the system relaxes completely before additional stress is added.

The BTW model[1] is the only model that is driven stochastically by randomly incrementing the stress of individual sites. Stress is added one unit at a time to a randomly chosen site. If the stress at that site is greater than or equal to a uniform threshold h_c, then the site loses four units of stress, and one unit of stress is given to each of its neighbors. On an $L \times L$ system the rules of evolution are given by:

$$\text{Driving}: \quad h(i,j) \to h(i,j) + 1, \quad i,j \in (1, L) \quad \text{random.}$$

$$
\begin{aligned}
\text{Toppling } (h(i,j) \geq h_c): \quad & h(i,j) \to h(i,j) - 4, \\
& h(i \pm 1, j) \to h(i \pm 1, j) + 1, . \\
& h(i, j \pm 1) \to h(i, j \pm 1) + 1
\end{aligned}
\tag{4}
$$

$$
\begin{aligned}
\text{Boundary Conditions}: \quad & h(i, L+1) = h(i, 0) = 0, \\
& h(L+1, j) = h(0, j) = 0.
\end{aligned}
$$

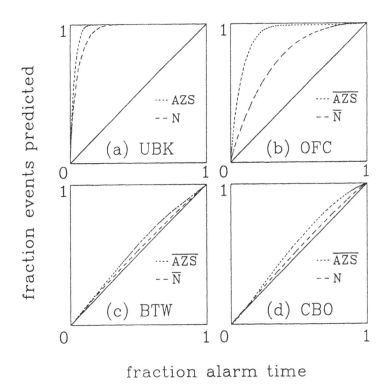

fraction events predicted

(a) UBK ···· AZS -- N

(b) OFC ···· $\overline{AZS}$ -- $\overline{N}$

(c) BTW ···· $\overline{AZS}$ -- $\overline{N}$

(d) CBO ···· $\overline{AZS}$ -- N

fraction alarm time

FIGURE 3 Success curves for the (a) UBK, (b) OFC, (c) BTW, and (d) CBO models. For each model, AZS (or $\overline{AZS}$) leads to more precise predictions than N (or $\overline{N}$). To obtain these results for each of the system, we have crudely optimized over the space-time windows. The spatial windows are the entire system for the SOC models, and 213 blocks for the UBK model. The time windows correspond to $\Delta t = .1$ for the UBK model, 33 net grains added for the BTW model, .15 net stress added per site for the OFC model, and .007 net stress added per site in the measurement of $\overline{AZS}$ and .015 net stress added per site in the measurement of N for the CBO model.

Note that stress is dissipated only at the boundary.

The OFC model[17] is the only model in which the internal (i.e. away from the boundary) dynamics is intrinsically dissipative. The system is driven deterministically by increasing the stress uniformly over the whole system. When a site reaches the threshold h_c, the stress of that site is set to zero, and the neighboring sites are all given a fixed fraction $\alpha < .25$ of the stress of the toppling site. On an $L \times L$ system the rules of evolution are given by:

Driving : $h(i,j) \rightarrow h(i,j) + \epsilon$ $(\epsilon \rightarrow 0)$, $\forall i, j \in (1, L)$ deterministic.

Toppling $(h(i,j) \geq h_c)$: $h(i,j) \rightarrow 0$,

$$h(i \pm 1, j) \rightarrow h(i \pm 1, j) + \alpha h(i,j),\,.$$
$$h(i, j \pm 1) \rightarrow h(i, j \pm 1) + \alpha h(i,j) \tag{5}$$

Boundary Conditions : $h(i, L + 1) = h(i, 0) = 0$,
$$h(L + 1, j) = h(0, j) = 0.$$

In this manner a fraction $(1 - 4\alpha)$ of the original stress is dissipated when any site in the interior of the system topples, and dissipation increases at the boundaries.

The CBO model[7] is the only model in which the relaxation dynamics explicitly takes place using long-range interactions, rather than by redistributing stress only to neighboring sites. In this case, (i,j) represents the location of a block on a two-dimensional square lattice, and $\mathbf{h}(i,j)$ has two components representing the stress in two out of the four springs joining the block at (i,j) to its neighbors. In particular, $h_x(i,j)$ represents the stress in the compressional spring joining the block at (i,j) to the block at $(i+1, j)$, and $h_y(i,j)$ represents the stress in the shear spring joining the block at (i,j) to the block at $(i, j + 1)$. The system is driven by deterministically increasing the stress levels in the shear springs, and when one of these spring exceeds its threshold $h_{c,y}(i,j)$, the spring breaks (the stress is set to zero, and the threshold for that site is reset to a randomly chosen value) and stress is redistributed throughout the lattice in a manner that is based on elastic dipole interactions. On an $L \times L$ system the rules of evolution are given by:

Driving : $h_y(i,j) \rightarrow h_y(i,j) + \epsilon$ $(\epsilon \rightarrow 0)$,   for all shear springs

Toppling $(h_r(i,j) \geq h_{c,r}(i,j))$: $h_r(i,j) \rightarrow 0$,

$$h_{c,r}(i,j) \rightarrow h'_{c,r}(i,j) \in [0,1], \quad r, s \in (x,y),$$
$$h_s(k,l) \rightarrow h_s(k,l) + h_r(i,j)G_{r,s}(i,j,k,l),$$

Boundary Conditions : $h_x(m,n) = h_y(m,n) = 0, \,\forall\, m, n < 1, \; m, n > L, \tag{6}$

where the toppling rule applies individually to both the x- and y-components of $\mathbf{h}(i,j)$. Here $\mathbf{G}(i,j,k,l)$ is defined in terms of the lattice dipole Green's functions which determine the change in stress for the springs at all other sites (k,l) when the shear or compressional spring at site (i,j) is broken. Because the interaction is long range, all events involve some dissipation, although events near the boundary tend to be more dissipative than those that are confined to the interior.

Unlike the UBK model, in the SOC models the distinction between small precursory events, and the large events that we attempt to predict, is no longer a sharp feature in the statistical distribution, and the largest events span essentially the entire system. Thus for the SOC models we define only one spatial region corresponding to the entire system, and attempt to predict events of a size s that is greater than or equal to the size $\tilde{s}$ where we estimate that finite size effects first become apparent (see Figure 1).

The results are illustrated in Figure 3, where we plot the success curves of the SOC models, along with our previous results for the UBK model. In each case both N and AZS show at least some correlation with a coming large event, and in most cases it is an anticorrelation, in which case the complement measures, i.e., lack of activity $\overline{N}$ (quiescence) and lack of active zone size $\overline{AZS}$, are shown in the figure. In these cases TIPs are turned on when our previous measures N and AZS take values below, rather than above, a specified threshold. Of the SOC models, the OFC model is clearly most predictable. Interestingly, again it is the spatial measure—in this case $\overline{AZS}$—that is most effective: 90% events predicted with alarm times of order 20%. Similarly, for the BTW and CBO models, $\overline{AZS}$ outperforms temporal precursors ($\overline{N}$ for the BTW model, and N for the CBO model). However, compared to the UBK and OFC models, the gain over purely random methods is significantly reduced.

In each case there is at least some correlation between small-scale activity and coming large events. The poor performance of the BTW and CBO models indicates that the correlations need not be strong, so that if the data were limited, it might be impossible to insure the statistical significance of these correlations (we have verified that our results have converged by comparing the success curves obtained for exponentially increasing catalog lengths). In the SOC models the precursors based on quiescence are typically most effective because, unlike the UBK model, the stress on a site is set to zero each time a block slips, independent of the event size. Thus a lack of events is more likely to signify that the system is near the slipping threshold. However, in both the UBK and SOC models, large events involve large spatial regions, so that in order for a large event to occur, the system must be near threshold across a relatively large region in space. In all of the models considered here, methods based on spatial rather than temporal precursors are most effective, because they provide the most direct measure of the development of such regions.

4. APPLICATIONS TO THE EARTH

In the earth certain tendencies towards clustering of earthquakes in space and time have been noted,[21] and are the phenomenological basis for the initial definition of the spatial regions of investigation in M8. However, the seven M8 precursors that are measured within each region are based only on the time series of events in the region, with no additional attention paid to the spatial locations of events. In this section we describe some of the results obtained when active zone size (AZS) is measured in the earth.[14] In particular, we compare the performance of the original M8 algorithm to a modified version in which measures based on AZS replace measures based on activity (N) in the algorithm. In the study we compare the results of the modified algorithm to the results obtained by Healy, Kossobokov, and Dewey in their Circum Pacific study.[10] Here we will restrict our attention to the portion of the test that applies to the Western United States, and the goal will be to predict events with magnitude greater than or equal to 7.5.

The comparison is based on data from the Earthquake Data Base System compiled at the National Earthquake Information Center (Golden, Colorado). It consists of the Global Hypocenter CD-ROM Data Base (Version 1.0) through 1988 and the Preliminary Determination of Epicenters (PDE) from 1989 up to the present. To insure completeness over the range of magnitudes considered, the catalog is restricted to earthquakes with magnitude M greater than or equal to 4.0 from 1963 on. In typical applications of M8 the catalog is sorted into main shocks and aftershocks using a specified declustering algorithm, although the measures based on AZS can be defined with or without a separation of aftershocks from main shocks in the catalog.[1]

In the models AZS was defined to be the total number of blocks that have slipped within a given time window. The most direct analogy in the earth would be made if one could somehow estimate the total area of rupture in a given region within a specified period of time, being careful not to count overlapping areas more than once. Of course, such precise information is not currently available except perhaps in very specific locations where microseismicity studies are underway. Furthermore, in the models the spatial regions of investigation that are considered are at most of order the size of the large event to be predicted. In comparison, the regions defined in M8 are an order of magnitude larger in radius than the size of the large earthquake, so that the correlations which contribute to a prediction are *a priori* more long range. For these reasons, in comparison to the models, when we define AZS in the earth it is necessarily a more coarse grained measure.

As shown in Figure 4, to define AZS we impose a grid on each of the spatial regions considered in M8. In this particular case, the grid is defined so that the side of each small box is 1/32 of the diameter of the circle. Then, for each earthquake of magnitude greater than a prescribed threshold that occurs in the region within the prescribed time window, we shade in the box that contains the epicenter of that earthquake. The measure AZS is thus defined to be the number of shaded boxes, normalized by the cumulative number of shaded boxes over a very long time period.

---

[1] A main shock is defined to be the largest event in a sequence of earthquakes, consisting of a main shock followed by aftershocks, and possibly preceded by foreshocks. Various different algorithms, referred to as declustering algorithms, exist for associating aftershocks with a particular main shock in terms of their relative separation in space and time from the main shock.

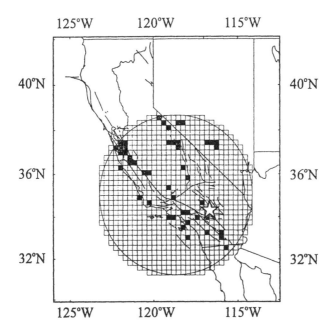

FIGURE 4 Definition of active zone size. Circle No. 116 from the Healy et al.[10] study is used to illustrate our definition of AZS. The circle is enclosed in a square grid of 32-by-32 smaller squares, measuring 26.6 km on a side. Grid boxes are shaded when the epicenter of at least one earthquake with magnitude greater than or equal to the cutoff falls in the square. The figure illustrates AZS evaluated over a six-year time window ending on January 1, 1990 (which corresponds to the time that the TIP is declared in the region). The location of the epicenter of the Landers 1992 earthquake is marked with a star.

This final normalization, while technically unnecessary, allows values of AZS from different regions (which may have different densities of active faults) to be more readily compared.

In M8 four of the seven measures are based directly on the activity. Two of the measures, $N1$ and $N2$, are simply the value of N for two different lower magnitude cutoffs. For $N1$ the cutoff is M10, the magnitude of events that occur at a rate of 10 per year on average. For $N2$ the cutoff is M20, the magnitude of events that occur at a rate of 20 per year on average. The other two activity measures are referred to as $L1$ and $L2$, and are defined to be the deviations of the current values of $N1$ and $N2$ from the long-term trend. Thus the L measures are sensitive to fluctuations in the activity.

To substitute AZS-based measures for N-based measures in M8, we define four analogous measures for the modified algorithm. We define $AZS1$ and $AZS2$, in direct analogy with $N1$ and $N2$, be be the value of AZS defined with respect to the lower magnitude cutoffs M10 and M20, respectively. The analogy between the other two AZS-based measures and the N-based measures they replace is necessarily somewhat less direct. Because the long time cumulative value of AZS saturates at a value representative of the fault density, an exact analogy with the L measures would differ from the original AZS measures only by a constant. Thus we define two measures $STZ1$ and $STZ2$ (for short-term zone) to correspond to AZS defined with a shorter time window of six months (and the same two magnitude cutoffs). Like the analogous N-based measures, these additional functions are sensitive to fluctuations in the value of the basic measure.

In Figure 5 we compare the time series of values of $N1$ and $N2$ to the corresponding values of $AZS1$ and $AZS2$ during the period leading up to the Landers earthquake in a circle of investigation which contains the epicenter. Each measure is updated every six months. The shaded data points illustrate those measurements which cast a vote in favor of issuing a TIP. The curves illustrated in Figure 5 only correspond to a small portion of the information that is compiled in M8. Nonetheless, they convey the basic observed behavior. The trends are typically weak, which suggests that neither of these measures could be used alone to make reliable forecasts. While these precursors are clearly correlated, AZS appears to be somewhat less volatile than N. With some effort it is possible to detect increases and drops in the values of the measures which persist over times periods of two to five years. In fact, one of the more striking features that is seen in Figure 5 is the drop in each of the measures which begins around 1990 and persists until the Landers earthquake in 1992. While decreases and minima were considered as candidate measures in the early development of M8, such measures were not seen to be effective in those early tests (it may be worth reconsidering these along with other possibilities on a larger statistical base). Note also that the AZS-based measures takes elevated values immediately *after* the Landers earthquake. This increase is associated with the broad regional activation involving much of the state of California which received widespread attention following the Landers earthquake.

Comparing the original and modified algorithms[14] we found that over the test period the AZS-based version of M8 slightly outperformed the original version both in terms of reduced alarm time, and a successful prediction of the Landers Earthquake in 1992, which was missed with the original algorithm. The modified algorithm also showed improvements over the original version when the methods were tested on smaller target events (so that the sizes of the circles were reduced and the goal was to predict events of magnitude greater than or equal to 7.0). Of course a much longer test must be conducted to obtain statistically significant results. Our main conclusion from the initial study is that consideration of alternate measures, and in particular measures based on spatial features, is likely to be worthwhile.

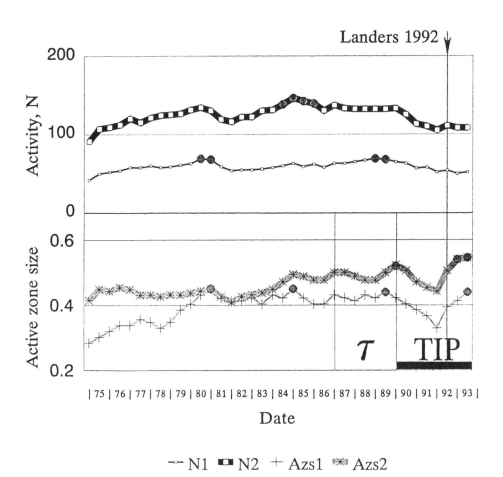

FIGURE 5 Activity (N) and active zone size (AZS) prior to the Landers 1992, $M = 7.6$ earthquake in circle 116 from the Healy et al.[10] study which contains the earthquake. Here $N1$ and $N2$ (and $AZS1$ and $AZS2$) are values of the measure corresponding to different lower magnitude cutoffs. The original version of M8 fails to predict the Landers earthquake in this circle. However, when AZS substitutes for N the earthquake is successfully predicted. The scale for N is shown on the left, while that for AZS is shown on the right. For each precursor, high (top 10%) values are marked with a shaded circle covering the data points, and the thresholds are determined within each circle and for each precursor individually on the basis of the values observed in 1975-1991. If six of the seven measures are high within a period τ of three years, a TIP is issued for a period of five years.

5. WHAT HAVE WE LEARNED?

Predicting earthquakes is intrinsically a difficult problem. Because the time scales over which tectonic stresses build are long, the amount of available data as well as the rate at which we can expect to gather more is severely limited. Under such circumstances, there is always a danger that methods which work well on the catalogs that are available today may not work well in the future—either due to statistical fluctuations or because the studies are inherently retroactive. To learn more about the processes that lead up to large earthquakes, it is useful to design forecasting methods as objectively and concretely as possible, and to conduct forward tests of these methods as is being done with M8. It is still too early to determine the ultimate promise and/or limitations of this method for the earth. For that reason it is especially useful to employ models as a complementary means of objective testing, as well as tools for algorithm development.

M8 is only one possible method of forecasting large earthquakes based on the statistics of small-scale seismicity. As more is learned about the successes and failures of the algorithm, it is natural to reevaluate aspects of the design. One of the important questions which this study begins to address is the selection of precursors. While such issues were considered by Keilis-Borok and Kossobokov in the initial formulation of M8, the measures that were ultimately chosen were selected in part for simplicity and in part for their performance on a relatively small data set. Now that larger catalogs are being considered,[10] it would be worthwhile to retest a variety of measures describing both spatial and temporal correlations in the catalog. Of course AZS is only one manner in which spatial information can be measured. Other possibilities include monitoring higher-order correlations in the relative distances between epicenters.

The more general question of accounting for precursory patterns in a complete and systematic way remains an open question in the representation of all kinds of complex spatiotemporal dynamical systems. In seismological applications, advances in this general area should ultimately lead to systematic modifications and improvements of M8 and related prediction techniques. In addition, we gain insight into the ultimate limitations in predictability of the underlying system given the incomplete description of the dynamics that is contained in the catalog of events. Such information will help to determine how well future generations of M8 can be expected to perform even in the best possible scenario. The relatively large fraction of time occupied by alarms in the present studies indicates that even the most optimistic evaluation of the current performance of the algorithm is not meeting the goals of intermediate-term prediction, where the target is to maintain alarms for only a few percent of the recurrence interval. Furthermore, even in the most predictable of the models, the UBK model, the best precursor we have found still requires alarms over 8% of the space-time volume to obtain a 90% success rate for prediction. On the other hand, in seismology perhaps the most pertinent issues still have less to do with the details of how long alarms must be maintained, and instead reduce to the question of whether earthquakes are predictable at all using these techniques. The answer to this question for the models is clearly affirmative. Determining the answer for real earthquakes is simply a matter of time.

The advantage in studying model systems is that optimization questions—whether they be questions of evaluating the best precursor, or determining the best way to select a set of precursors—do in principle have decidable answers right now. In some cases the optimization question may be complex and very high dimensional, but it is certainly well defined. In contrast, for the seismic catalogs because of the limited amount of data which is available it is impossible to conclude with certainty that there is a statistically significant correlation between precursory patterns that are measured and large events. In the current studies results which incorporate a relatively large number of target events are obtained by combining data for many circles of investigation. Thus for seismicity studies perhaps the most relevant results we can obtain using models are those that estimate the effects of limited information on the predictions that are made, or more precisely at what minimum temporal catalog length a spatial average becomes equivalent to a long time average. This question was addressed by Pepke et al.[18] for the UBK model, where it was found that there were large fluctuations in the success rate of the algorithm for a combined set of many shorter catalogs; however, the performance converged to its long time value after roughly one recurrence interval worth of data was obtained. This suggests (but is a long way from proving) that the time scale over which improvements might be expected is certainly appreciable (of order the seismic cycle) but that perhaps one need not wait unreasonably long times to obtain substantial gains.

Finally, it is worth emphasizing that the detailed dynamics and correlations between events associated with each of the models fails to capture some important aspects of the phenomena that are observed in the earth. One of the most striking disparities is the lack of aftershocks in all of the models. This has an impact on our ability to study M8, since one of the seven measures defined in M8 is based solely on aftershocks. Secondly, the fact that the circles of investigation in M8 are taken to be an order of magnitude larger than the estimated length of the target events, while in the models the regions are defined to be at most the size of the rupture to be predicted is indicative of the existence of some long-range correlations in the earth which are not represented yet in models. The development of more realistic models in the future may remedy these differences.

ACKNOWLEDGMENTS

This work was supported by the David and Lucile Packard Foundation, and NSF grants DMR-9212396 and PHY89-04035, and an INCOR grant from LANL.

REFERENCES

1. Bak, P., C. Tang, and K. Wiesenfeld. *Phys. Rev. Lett.* **59** (1987): 381.
2. Burridge, R. and L. Knopoff. "Model and Theoretical Seismicity." *Bull. Seism. Soc. Am.* **57** (1967): 341–371.
3. Carlson, J., and J. Langer. "Properties of Earthquakes Generated by Fault Dynamics." *Phys. Rev. Lett.* **62** (1989): 2632.
4. Carlson, J., and J. Langer. *Phys. Rev. A* **40** (1989): 6470.
5. Carlson, J. "A Two-Dimensional Model of a Fault." *Phys. Rev. A* **44** (1991): 6226. As seen in this paper, the 2-D UBK model differs little from the 1-D case.
6. Carlson, J., J. Langer, B. Shaw, and C. Tang. "Intrinsic Properties of a Burridge-Knopoff Model of an Earthquake Fault." *Phys. Rev. A* **44** (1991): 884.
7. Chen, K., P. Bak, and S. Obukhov. "Self-Organized Criticality in a Crack-Propagation Model of Earthquakes." *Phys. Rev. A* **43** (1991): 625.
8. Gutenberg, B., and C. Richter. *Seismicity of the Earth and Related Phenomena.* Princeton: Princeton University Press, 1954.
9. Habermann, R. E. "Seismicity Rate Variations and Systematic Changes in Magnitudes in Teleseismic Catalogs." *Tectonophysics* **193** (1991): 277.
10. Healy, J., V. Kossobokov, and J. Dewey. *U.S. Geol. Surv. Open File Rep.* (1992): 92–401.
11. Jones, L. M. "Foreshocks and Time-Dependent Earthquake Hazard Assessment in Southern California." *Bull. Seism. Soc. Am.* **75** (1985): 1669–1679.
12. Keilis-Borok, V. I., and I. M. Rotwain. "Times of Increased Probability of Strong Earthquakes (M ≥ 7.5) Diagnosed by Algorithm M8 in Japan and Adjacent Territories." *Phys. Earth & Planet. Int.* **61** (1990): 57–72.
13. Keilis-Borok, V. I., and V. G. Kossobokov. *Phys. Earth & Planet. Int.* **61** (1990): 73–83.
14. Kossobokov, V. G., and J. M. Carlson. "Active Zone Size Versus Activity: A Study of Different Seismicity Patterns in the Context of the Prediction Algorithm M8." *J. Geophys. Res.* **100** (1995): 6431–6441.
15. Minster, J.-B.,and N. P. Williams. *EOS Transactions AGU 1992 Fall Meeting* **73** (1992): 366.
16. Molchan, G. M. "Structure of Optimal Strategies in Earthquake Prediction." *Tectonophysics* **193** (1991): 267.
17. Olami, Z., H. Feder, and K. Christensen. "Self-Organized Criticality in a Continuous, Nonconservative Cellular Automaton Modeling Earthquakes." *Phys. Rev. Lett.* **68** (1992): 1244.
18. Pepke, S. L., J. M. Carlson, and B. E. Shaw. *J. Geophys. Res.* **99** (1994): 6769.

19. Pepke, S. L., and J. M. Carlson. "Predictability of Self-Organizing Systems." *Phys. Rev. E* **50** (1994): 236.

20. Shaw, B., J. Carlson, and J. Langer. "Patterns of Seismic Activity Preceding Large Earthquakes." *J. Geophys. Res.* **97** (1992): 479.

21. Scholz, C. *The Mechanics of Earthquakes and Faulting.* New York: Cambridge University Press, 1990.

22. Working Group on California Earthquake Prediction. *U.S. Geol. Surv. Open File Rep.* (1988): 88–398.

Index